Engineering Optimization: Methods and Applications

Optimization carries great significance in both human affairs and the laws of nature. It refers to a positive and intrinsically human concept of minimization or maximization to achieve the best or most favorable outcome from a given situation. Besides, as the resources are becoming scarce there is a need to develop methods and techniques which will make the systems extract maximum from minimum use of these resources, i.e. maximum utilization of available resources with minimum investment or cost of any kind. The resources could be any, such as land, materials, machines, personnel, skills, time, etc. The disciplines such as mechanical, civil, electrical, chemical, computer engineering as well as the interdisciplinary streams such as automobile, structural, biomedical, industrial, environmental engineering, etc. involve in applying scientific approaches and techniques in designing and developing efficient systems to get the optimum and desired output. The multifaceted processes involved are designing, manufacturing, operations, inspection and testing, forecasting, scheduling, costing, networking, reliability enhancement, etc. There are several deterministic and approximation-based optimization methods that have been developed by the researchers, such as branch-and-bound techniques, simplex methods, approximation and Artificial Intelligence-based methods such as evolutionary methods, Swarm-based methods, physics-based methods, socio-inspired methods, etc. The associated examples are Genetic Algorithms, Differential Evolution, Ant Colony Optimization, Particle Swarm Optimization, Artificial Bee Colony, Grey Wolf Optimizer, Political Optimizer, Cohort Intelligence, League Championship Algorithm, etc. These techniques have certain advantages and limitations and their performance significantly varies when dealing with a certain class of problems including continuous, discrete, and combinatorial domains, hard and soft constrained problems, problems with static and dynamic in nature, optimal control, and different types of linear and nonlinear problems, etc. There are several problem-specific heuristic methods are also existing in the literature.

This series aims to provide a platform for a broad discussion on the development of novel optimization methods, modifications over the existing methods including hybridization of the existing methods as well as applying existing optimization methods for solving a variety of problems from engineering streams. This series publishes authored and edited books, monographs, and textbooks. The series will serve as an authoritative source for a broad audience of individuals involved in research and product development and will be of value to researchers and advanced undergraduate and graduate students in engineering optimization methods and associated applications.

Sudath Rohan Munasinghe

Optimization and Optimal Control in a Nutshell

Springer

Sudath Rohan Munasinghe
Department of Electronic
and Telecommunication Engineering
University of Moratuwa
Moratuwa, Sri Lanka

ISSN 2731-4049 ISSN 2731-4057 (electronic)
Engineering Optimization: Methods and Applications
ISBN 978-981-97-8166-9 ISBN 978-981-97-8167-6 (eBook)
https://doi.org/10.1007/978-981-97-8167-6

This Springer imprint is published by the registered company Springer Nature Singapore Pte Ltd.
The registered company address is: 152 Beach Road, #21-01/04 Gateway East, Singapore 189721, Singapore

This book is dedicated to my wife, two daughters, and my son for their patience and caring support extended.

Acknowledgement

The author sincerely acknowledges the support and guidance provided by Prof. Lori Leonard, Prof. Ronnie Coffman, and Prof. Richard Cahoon of the Department of Global Development, CALS, Cornell University, Ithaca, New York, USA.

Contents

About the Author

Dr. Sudath Rohan Munasinghe is a Professor of Control Systems and Robotics at the Department of Electronic and Telecommunication Engineering of the University of Moratuwa, Sri Lanka. He is also a visiting Scientist at the Department of Global Development of the College of Agriculture and Life Sciences of Cornell University, USA. He teaches courses at undergraduate and graduate levels in the areas of control systems, robotics, unmanned aerial vehicles, and learning-based controls. His research spans several areas including vision-force sensor fusion, robot manipulator control, mobile robot self-navigation, aerial manipulation, non-invasive fetal monitoring, and elephant rumble detection using machine learning. Dr. Munasinghe was a 2021 Fulbright scholar.

List of Figures

Chapter 1
Basic Optimization Concepts

Optimization aims at finding a unique point, a unique state, or a unique function (a sequence of points or states) that is associated with the minimum cost or maximum benefit with respect to a certain criterion, subject to constraints. Function optimization is largely based on calculus developed by Sir Isaac Newton and Gottfried Wilhelm Leibniz. The most comprehensible optimization process is finding the minima or maxima of a single variable function $y = f(x)$ by setting the first derivative $\frac{dy}{dx} = 0$, which can be extended to multi-variable functions $y = f(x_1, x_2, \dots x_n)$ by setting all partial derivatives to zero $\frac{\partial y}{\partial x_i} = 0; i = 1, 2, \dots n$. More often, the functions to be optimized are costs that can be modeled in the quadratic form $f(\mathbf{x}) = \mathbf{x}^\mathrm{T}\mathbf{Q}\mathbf{x}$, which allows wider applicability of mathematical tools in the optimization process, hence quadratic optimization.

Optimization is sometimes unconstrained, meaning that any value for x in a single variable case and any values for $x_1, x_2, \dots x_n$ in a multi-variable case are acceptable as solutions. However, this is not the case in most of the real-world applications. When realistic constraints $g_i(x_1, x_2, \dots x_n) \leq 0; i = 1, 2, \dots m$ exist, the best solution that satisfies such constraints is to be found, hence constrained optimization.

In some optimization problems, the unconstrained solution satisfies the constraints, in which case the constraints are redundant. However, when the unconstrained solution violates the constraints, the next best solution that satisfies the constraints is selected. In the latter case, at the optimal solution, the function gradient $\nabla f = [\frac{\partial f}{\partial x_1}, [\frac{\partial f}{\partial x_2} \dots [\frac{\partial f}{\partial x_n}]^T$ and the gradient of the constraints $\nabla g_i = [\frac{\partial g_i}{\partial x_1}, [\frac{\partial g_i}{\partial x_2} \dots [\frac{\partial g_i}{\partial x_n}]^T; i = 1.2 \dots m$ are aligned. Joseph-Louis Lagrange used this gradient alignment with the Lagrange multipliers to build the condition for constrained optimization.

S. R. Munasinghe, *Optimization and Optimal Control in a Nutshell*, Engineering Optimization: Methods and Applications,
https://doi.org/10.1007/978-981-97-8167-6_1

To find the optimum function (path) between two states involves cost function minimization over all possible functions between the same two end-states. In this problem, the functions are defined using a parameter (ϵ), and the cost of each function is determined by an integral that is evaluated between the two end-states. The optimum function is determined by differentiating the cost integral with respect to the function parameter ϵ. This thinking was inspired by two problems raised in the mathematical community: the minimal resistance problem by Sir Isaac Newton in 1687 and the Brachistocrone problem by Johann Bernoulli in 1696. The solution process is now known as functional optimization, and the section of mathematics involved is the calculus of variations, as defined by Leonhard Euler and Joseph-Louis Lagrange in 1756.

Functional optimization can be performed using cost functions based on either Lagrangian L or Hamiltonian H, where $H = L + \boldsymbol{\lambda}^T \mathbf{f}$. A set of conditions for optimality is obtained for each of the two approaches. It is known that the Hamiltonian approach is a bit more efficient than the Lagrangian approach. When it comes to the optimization of control problems, the control effort u is also included in the cost function that is formed using Lagrangian or Hamiltonian, and the rest of the optimization process remains unchanged. However, the optimization process being an offline process, the control sequence naturally becomes an open-loop control process $u = f(t)$, which is not acceptable for most of the control problems in the real world. To develop closed-loop optimal control, it is required to determine the control signal as a function of the state $u - f(\mathbf{x})$, which is possible if the cost function is linear quadratic $\frac{1}{2}\int \left\{\mathbf{x}(t)^T Q(t)\mathbf{x}(t)\right\} dt$. This leads to the development of a well-known linear quadratic regulator (LQR).

In some control problems, the objective is to maximize the outcome while minimizing the cost $\max\{\Gamma - \int L(\mathbf{x}, \mathbf{u})\}dt$. More often, the maximum outcome is related to the maximum control signal. The maximum principle by Lev Pontryagin in 1956 explains how the concept is implemented using Lagrange or Hamiltonian-based objective functions. As expected, the control signal under the maximum principle is either maximum or zero (minimum), and no other control level is used within the period of control. Therefore, aggressive control switching is obvious, hence the name Bang–Bang control.

Most of the control systems in the modern world are discrete-time; hence, it is required to develop optimal controls based on the discrete-time state-space models $\mathbf{x}(k+1) = \mathbf{A}\mathbf{x}(k) + \mathbf{B}\mathbf{u}(k)$. Because all future states are deterministic given the initial state and successive controls, for a given initial state, the optimal control sequence that minimizes a quadratic cost $\mathbf{x}(k)^T \mathbf{Q}\mathbf{x}(k) + \boldsymbol{u}(k)^T \mathbf{R}\boldsymbol{u}(k)$ can be determined in an offline manner as $\mathbf{u}^* = -\bar{\mathbf{H}}^{-1}\bar{\mathbf{F}}\mathbf{x}(0)$. Yet again, offline calculation of control sequences is hardly accepted in real-world applications due to model uncertainty, disturbances, and noise. Hence, the formulation of a feedback form for the implementation of discrete-time optimal control is necessary. The approach is simple: we may continue to use the offline determination of the optimal control sequence for a finite time into the future but implement only the immediate control signal, and the process is repeated at each step based on the actual states. This concept was first proposed and implemented by C. R. Cutler and B. L. Ramaker in 1980, and it is now widely known as model predictive control (MPC).

The next chapter presents how to find the optimum point in scalar and vector functions. It also demonstrates the process for a quadratic function.

Chapter 2
Unconstrained Function Optimization

This chapter presents the necessary and sufficient conditions for an optimum point in a scalar function. The same is then extended to vector functions. The determination of the optimum point in a quadratic function is presented next. This chapter does not consider constraints.

2.1 Minima of a Scalar Function

A scalar function $f(x) \in \mathbb{R}$ is shown in Fig. 2.1, where the minimum point is indicated as x^* given by the following expression.

$$x^* = \min_x f(x) \tag{2.1}$$

It is assumed that $f(x)$ is smooth and continuously differentiable, i.e., $\frac{df(x)}{dx}$, $\frac{d^2 f(x)}{dx^2}$...exist. Around the locality ϵ of the stationary point, the function satisfies the following condition:

$$f(x^*) < f(x + \alpha); \, |\alpha| < \epsilon \tag{2.2}$$

The function shown in Fig. 2.1 shows multiple minimum points. Hence, it is critically important that the search for the minima start around x^*, the global minima. However, if the function $f(x)$ is convex, as shown in Fig. 2.2, $f(x)$ has only one minima, and the search can start from any arbitrary point. Convex functions have the property $f(\alpha a + (1 - \alpha)b) \leq \alpha f(a) + (1 - \alpha) f(b)$; $\alpha \in [0, 1]$ for any two points, $x = a$ and $x = b$.

S. R. Munasinghe, *Optimization and Optimal Control in a Nutshell*, Engineering Optimization: Methods and Applications,
https://doi.org/10.1007/978-981-97-8167-6_2

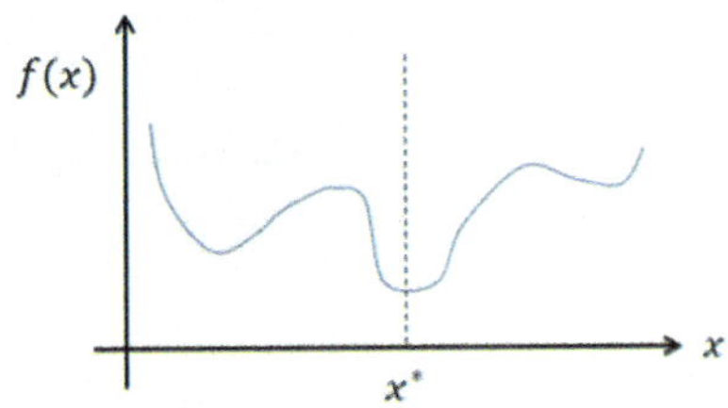

Fig. 2.1 A Scalar function $f(x)$

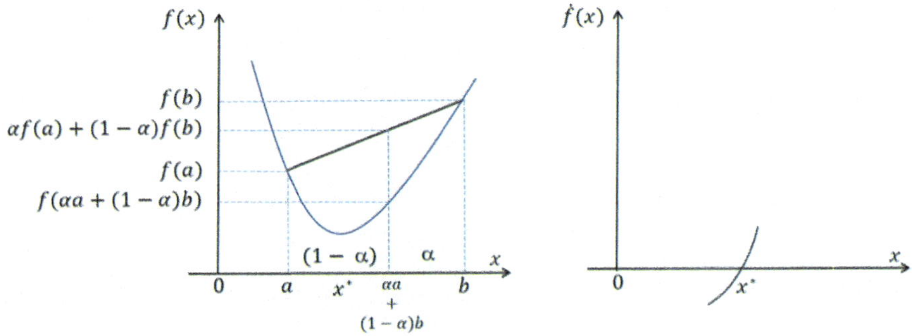

Fig. 2.2 A convex function (left) and the variation of the first derivative of a convex function across the stationary point x^* (right)

2.1.1 Necessary Condition for Optimality

The condition for optimality, also known as the first-order necessary condition (FONC), is as follows:

$$\dot{f}(x) = 0 \tag{2.3}$$

2.1.2 Sufficient Condition for Minima

When the stationary point is a minima, the slope changes from negative ($\dot{f}(x) < 0; x < x^*$) to positive ($\dot{f}(x) > 0; x > x^*$) across x^* as shown in Fig. 2.2(right), hence the sufficient condition for a minima is as follows:

$$\ddot{f}(x)|_{x^*} > 0 \tag{2.4}$$

2.2 Minima of a Vector Function

A convex vector function of two variables x_1, x_2 is shown in Fig. 2.3.

2.2.1 Necessary Condition for Optimality

It is obvious that, at the stationary point, the function needs to have zero gradients with respect to both variables x_1, x_2. This condition is generalized to a vector function of n variables, and the point x^* is a stationary point if the following first-order necessary condition is satisfied.

$$\nabla_{\mathbf{x}} f(\mathbf{x})|_{x^*} = \begin{bmatrix} \frac{\partial f(\mathbf{x})}{\partial x_1} \\ \frac{\partial f(\mathbf{x})}{\partial x_2} \\ \vdots \\ \frac{\partial f(\mathbf{x})}{\partial x_n} \end{bmatrix}_{x^*} = 0 \tag{2.5}$$

The vector $\nabla_{\mathbf{x}} f(\mathbf{x})$ is called the Jacobian.

2.2.2 Sufficient Condition for Minima

Following the argument in (2.4), at the minima of a vector function, the second partial derivatives of the function with respect to all the variables are positive. Hence, the necessary condition for the stationary point x^* to be a minima is as follows:

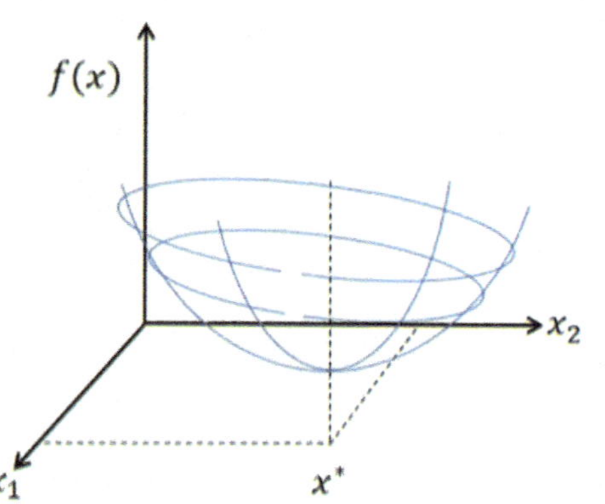

Fig. 2.3 A scalar function $f(x)$

$$\nabla_{\mathbf{x}}^2 f(\mathbf{x})|_{x^*} = \begin{bmatrix} \frac{\partial^2 f(\mathbf{x})}{\partial x_1^2} & \frac{\partial^2 f(\mathbf{x})}{\partial x_1 \partial x_2} & \cdots & \frac{\partial^2 f(\mathbf{x})}{\partial x_1 \partial x_n} \\ \frac{\partial^2 f(\mathbf{x})}{\partial x_2 \partial x_1} & \frac{\partial^2 f(\mathbf{x})}{\partial x_2^2} & \cdots & \frac{\partial^2 f(\mathbf{x})}{\partial x_2 \partial x_n} \\ \vdots & \vdots & \cdots & \vdots \\ \frac{\partial^2 f(\mathbf{x})}{\partial x_n \partial x_1} & \frac{\partial^2 f(\mathbf{x})}{\partial x_n \partial x_2} & \cdots & \frac{\partial^2 f(\mathbf{x})}{\partial x_n \partial x_n} \end{bmatrix}_{x^*} > 0 \tag{2.6}$$

The matrix $\nabla_{\mathbf{x}}^2 f(\mathbf{x})$ is called the Hessian. The above condition $\nabla_{\mathbf{x}}^2 f(\mathbf{x}) > 0$ is satisfied when the Hessian is strictly positive definite, which is verified by showing that all the eigenvalues of $\nabla_{\mathbf{x}}^2 f(\mathbf{x})$ are positive.

Example:
Minimize the cost function $f(x) = (x_1 - x_2^2)(x_1 - 3x_2^2)$.

Solution:
The given cost function is

$$f(x_1, x_2) = x_1^2 - 4x_1x_2^2 + 3x_2^4$$

The first-order necessary condition (FONC) for optimality is as follows:

$$\nabla_{\mathbf{x}} f(\mathbf{x})_{x^*} = \begin{bmatrix} \frac{\partial f(\mathbf{x})}{\partial x_1} \\ \frac{\partial f(\mathbf{x})}{\partial x_2} \end{bmatrix}_{x^*} = \begin{bmatrix} 2x_1 - 4x_2^2 \\ -8x_1x_2 + 12x_2^3 \end{bmatrix} = 0$$

The stationary point is $x^* = (0, 0)$. To check the nature of the stationary point (minima, maxima, or saddle point), let's check the Hessian as follows:

$$\nabla_{\mathbf{x}}^2 f(\mathbf{x})_{x^*} = \begin{bmatrix} \frac{\partial^2 f(\mathbf{x})}{\partial x_1^2} & \frac{\partial^2 f(\mathbf{x})}{\partial x_1 \partial x_2} \\ \frac{\partial^2 f(\mathbf{x})}{\partial x_2 \partial x_1} & \frac{\partial^2 f(\mathbf{x})}{\partial x_2^2} \end{bmatrix}_{x^*} = \begin{bmatrix} 2 & -8x_2 \\ -8x_2 & 36x_2^2 \end{bmatrix}_{x^*} = \begin{bmatrix} 2 & 0 \\ 0 & 0 \end{bmatrix} \geq 0$$

The Hessian is not strictly positive. The two eigenvalues are 2 and 0. Hence, the stationary point (0, 0) is neither a minima nor a maxima. It's a saddle point.

Eigenvalues of a Square Matrix

To determine the eigenvalues $\lambda_i; i = 1, 2..n$ of a $n \times n$ square matrix, solve the equatons $|A - \lambda I| = 0$. For example, the eigenvalues λ of matrix A

$$A = \begin{pmatrix} 1 & 2 \\ 3 & 4 \end{pmatrix}$$

are given by

$$|A - \lambda I| = 0$$

$$\left| \begin{pmatrix} 1 & 2 \\ 3 & 4 \end{pmatrix} - \begin{pmatrix} \lambda & 0 \\ 0 & \lambda \end{pmatrix} \right| = 0$$

$$\begin{vmatrix} 1-\lambda & 2 \\ 3 & 4-\lambda \end{vmatrix} = 0$$

$$(1-\lambda)(4-\lambda) - 6 = 0$$

$$\lambda^2 - 5\lambda - 2 = 0$$

$$\lambda = \frac{5 \pm \sqrt{(-5)^2 - 4(-2)}}{2}$$

$$\lambda = \frac{5 \pm \sqrt{33}}{2}$$

$$\lambda = 5.37 \text{ and } -0.37$$

If the matrix A is diagonal, then the eigenvalues are the diagonal elements.

2.3 Optimality of a Quadratic Function

Quadratic cost functions shown below are very commonly found in the optimal control field.

$$f(\mathbf{x}) = \frac{\mathbf{1}}{\mathbf{2}}\mathbf{x}^{\mathbf{T}}\mathbf{Q}\mathbf{x} + \mathbf{r}^{\mathbf{T}}\mathbf{x} + \mathrm{c} \tag{2.7}$$

where $\mathbf{Q}$ is a positive definite square matrix, $\mathbf{r}$ is a $n \times 1$ vector, and c is a constant. For example, if $\mathbf{x} = (x_1\ x_2)^T$ and $\mathbf{Q} = \begin{bmatrix} 1 & 0 \\ 0 & 1 \end{bmatrix}$ the expression $\mathbf{x}^T\mathbf{Q}\mathbf{x} = x_1^2 + x_2^2$ correctly represents the energy or cost associated with the variables. The term $\mathbf{r}^T\mathbf{x}$ represents the cost (if any) associated with the state $\mathbf{x}$, and the constant c is there to represent if there is any fixed cost. To determine $\nabla_{\mathbf{x}} f(\mathbf{x})$, let's use the following equation:

$$\begin{aligned} df &= [\nabla_{\mathbf{x}} f(\mathbf{x})]^T d\mathbf{x} \qquad (2.8) \\ &= \left[\nabla_{\mathbf{x}}\left(\frac{1}{2}\mathbf{x}^T\mathbf{Q}\mathbf{x} + \mathbf{r}^T\mathbf{x}\right)\right]^T d\mathbf{x} \\ &= \frac{1}{2}\left(d\mathbf{x}^T\mathbf{Q}\mathbf{x} + \mathbf{x}^T\mathbf{Q}d\mathbf{x}\right) + \mathbf{r}^T d\mathbf{x} \\ &= \frac{1}{2}\left((d\mathbf{x}^T\mathbf{Q}\mathbf{x})^T + \mathbf{x}^T\mathbf{Q}d\mathbf{x}\right) + \mathbf{r}^T d\mathbf{x} \\ &= \frac{1}{2}\left(\mathbf{x}^T\mathbf{Q}^T d\mathbf{x} + \mathbf{x}^T\mathbf{Q}d\mathbf{x}\right) + \mathbf{r}^T d\mathbf{x} \\ &= \mathbf{x}^T\left(\frac{\mathbf{Q}^T + \mathbf{Q}}{2}\right) d\mathbf{x} + \mathbf{r}^T d\mathbf{x} \end{aligned}$$

$$df = \left(\mathbf{x}^T\frac{(\mathbf{Q}^T + \mathbf{Q})}{2} + \mathbf{r}^T\right) d\mathbf{x} \tag{2.9}$$

Comparing (2.8) and (2.9)

$$\nabla_{\mathbf{x}} f(\mathbf{x}) = \left(\mathbf{x}^T \frac{(\mathbf{Q}^T + \mathbf{Q})}{2} + \mathbf{r}^T\right)^T \tag{2.10}$$

The $\mathbf{Q}$ matrix is often a diagonal matrix, hence $\mathbf{Q}^T = \mathbf{Q}$. In such situations, the Jacobian would take the following form.

$$\begin{aligned}\nabla_{\mathbf{x}} f(\mathbf{x}) &= \left(\mathbf{x}^T\mathbf{Q} + \mathbf{r}^T\right)^T \\ &= \mathbf{Q}^T\mathbf{x} + \mathbf{r} \\ &= \mathbf{Q}\mathbf{x} + \mathbf{r}\end{aligned} \tag{2.11}$$

Then, the first-order necessary condition (FONC) for optimality is as follows:

$$\nabla_{\mathbf{x}} f(\mathbf{x}) = \mathbf{Q}\mathbf{x} + \mathbf{r} = 0 \tag{2.12}$$

Hence, the optimal point is determined as follows:

$$\mathbf{x}^* = -\mathbf{Q}^{-1}\mathbf{r} \tag{2.13}$$

To check the nature of the optimal point, whether $\mathbf{x}^*$ is a minima, maxima, or saddle point, the Hessian of the quadratic cost is obtained by differentiating (2.11).

$$\nabla^2_{\mathbf{x}} f(\mathbf{x}) = \mathbf{Q} \tag{2.14}$$

When the $\mathbf{Q}$ matrix is positive definite ($\mathbf{Q} > 0$), the Hessian is positive $\nabla^2_{\mathbf{x}} f(\mathbf{x}) > 0$, hence the optimal point is a minima.

2.3.1 Quadratic Function Optimization

Exercise:
Find the optimum value of the cost $f(x_1, x_2) = x_1^2 + 2x_2^2 + 2x_1 - x_2 + 3$.

Solution:
The cost function can be written in the format of (2.7) as follows:

$$\begin{aligned}f(x_1, x_2) &= \frac{1}{2}\begin{bmatrix}x_1 & x_2\end{bmatrix}\begin{bmatrix}2 & 0 \\ 0 & 4\end{bmatrix}\begin{bmatrix}x_1 \\ x_2\end{bmatrix} + \begin{bmatrix}2 & -1\end{bmatrix}\begin{bmatrix}x_1 \\ x_2\end{bmatrix} + 3 \\ &= \frac{1}{2}\mathbf{x}^T\mathbf{Q}\mathbf{x} + \mathbf{r}^T\mathbf{x} + c\end{aligned}$$

This is a quadratic function with a positive definite weighting matrix ($\mathbf{Q} > 0$), hence the function has a minima. Then, from (2.13), the minima is

$$
\begin{aligned}
\mathbf{x}^* &= -\mathbf{Q}^{-1}\mathbf{r} \\
&= -\begin{bmatrix} 2 & 0 \\ 0 & 4 \end{bmatrix}^{-1} \begin{bmatrix} 2 \\ -1 \end{bmatrix} \\
&= \begin{bmatrix} -1 \\ 0.25 \end{bmatrix}
\end{aligned}
$$

The Matlab code in Appendix B carries out the optimization process. The result in Fig. 2.4 shows the cost contours on the x_1, x_2 plane and the cost surface. In both illustrations, the optimum point is indicated by a dot. The constant term s is not considered in the optimization process; hence, it should be added at the end; otherwise, the optimum point will stay above or below the cost surface. It is quite obvious that the optimum point is a minima.

The same result could be achieved using Jacobian as follows:

$$
\nabla_{\mathbf{x}} f(\mathbf{x}^*) = \begin{pmatrix} 2x_1^* + 2 \\ 4x_2^* - 1 \end{pmatrix} = 0
$$

$$
\mathbf{x}^* = \begin{pmatrix} -1 \\ 0.25 \end{pmatrix}. \tag{2.15}
$$

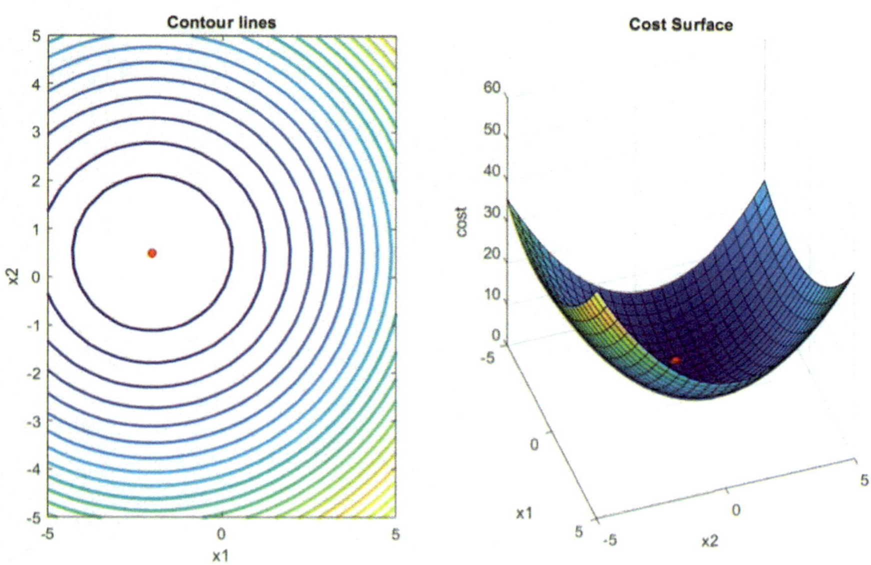

Fig. 2.4 Quadratic cost contour lines and the optimum (minimum) point

And the Hessian

$$\nabla_{\mathbf{x}}^2 f(\mathbf{x}) = \mathbf{Q} = \begin{bmatrix} 2 & 0 \\ 0 & 4 \end{bmatrix} > 0$$

being strictly positive definite, the optimality is a minima.

The next chapter presents how to incorporate constraints into the optimization process. It also demonstrates how the two Matlab commands `quadprog` and `fmincon` could be used in function optimization with constraints and boundary limits.

Chapter 3
Constrained Function Optimization

This chapter presents how to combine the equality constraints with the objective function in the optimization process. How to deal with inequality constraints using KKT conditions is presented next. Then, the implementation of quadratic optimization with equality constraints and boundary limits is demonstrated using the `quadprog` command in Matlab. In the end, the optimization of an arbitrary function subject to nonlinear constraints, linear constraints, and boundary limits is demonstrated using the `fmincon` command in Matlab.

3.1 Equality Constraints

Function optimization is often subject to constraints in the domain variables. Constraints are of two types: equality constraints and inequality constraints, and the constraints can be linear or nonlinear combinations of domain variables. The constrained optimization problem is defined as follows:

$$\min_{\mathbf{x}} f(\mathbf{x})$$

subject to

$$g_i(\mathbf{x}) = 0; i = 1, 2...m$$

where $g_i(\mathbf{x}) = 0$ are the equality constraints to be satisfied. The global optimum of the function $f(\mathbf{x})$ as shown in the previous sections without constraints might not be feasible when constraints are present. To visualize the constrained optimization problem, let's consider optimizing a paraboloidal cost function $f(\mathbf{x})$ of two variables

S. R. Munasinghe, *Optimization and Optimal Control in a Nutshell*, Engineering Optimization: Methods and Applications,
https://doi.org/10.1007/978-981-97-8167-6_3

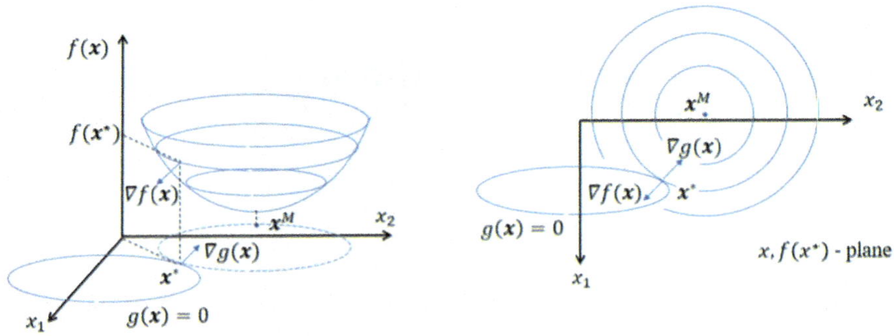

Fig. 3.1 Side view (left) and top view (right) of the cost function f(**x**) and the constraint g(**x**) = 0

(x_1, x_2) subject to an elliptical constraint $g(\mathbf{x}) = 0$ as shown in Fig. 3.1. The global minima x^M is not feasible because it doesn't satisfy the constraint $g(\mathbf{x}) = 0$; hence, the contours of higher cost are considered starting from x^M. The second contour is the one that touches the constraint for the first time; hence, the point x^* is the feasible minima subject to the constraint. As shown, the gradient of the constraint $\nabla g(\mathbf{x}) \in \mathbb{R}^2$ at x^* is normal to the constraint $g(\mathbf{x})$, and at the same point, the gradient of the cost function $\nabla f(\mathbf{x}) \in \mathbb{R}^2$ is normal to the cost contour as shown. Both gradients are defined on the (x_1, x_2) plane and share the point x^*. As shown, the two tangents of the two touching curves at x^* are colinear, and the gradients of the two curves at x^* are colinear as well. The two gradients are related as $\nabla f(\mathbf{x}) = -\lambda \nabla g(\mathbf{x})$. Then, the first-order necessary condition for constrained optimality is stated as follows:

$$\nabla_{\mathbf{x}} f(\mathbf{x}) + \lambda \nabla_{\mathbf{x}} g(\mathbf{x}) = \mathbf{0} \tag{3.1}$$

The constant λ in (3.1) is called the Lagrange multiplier. The above condition is formally written using the Lagrange function $\mathcal{L}$ as follows:

$$\mathcal{L}(\mathbf{x}, \lambda) = f(\mathbf{x}) + \lambda g(\mathbf{x}) \tag{3.2}$$

$$\nabla_{\mathbf{x}} \mathcal{L}(\mathbf{x}, \lambda) = \nabla_{\mathbf{x}} f(\mathbf{x}) + \lambda \nabla_{\mathbf{x}} g(\mathbf{x}) = \mathbf{0} \tag{3.3}$$

Example

Optimize $f(\mathbf{x}) = x_1x_2 + x_2x_3 + x_1x_3$ subject to $g(\mathbf{x}) = x_1 + x_2 + x_3 - 3 = 0$.

Solution

From (3.2) the Lagrangian is as follows:

$$\mathcal{L}(\mathbf{x}, \lambda) = x_1x_2 + x_2x_3 + x_1x_3 + \lambda(x_1 + x_2 + x_3 - 3)$$

From (3.3), the gradient of the Lagrangian is

$$\nabla_{\mathbf{x},\lambda}\mathcal{L} = \begin{bmatrix} \nabla_{x_1}\mathcal{L} \\ \nabla_{x_2}\mathcal{L} \\ \nabla_{x_3}\mathcal{L} \\ \nabla_{\lambda}\mathcal{L} \end{bmatrix} = \begin{bmatrix} x_2 + x_3 + \lambda \\ x_1 + x_3 + \lambda \\ x_2 + x_1 + \lambda \\ x_1 + x_2 + x_3 - 3 \end{bmatrix} = 0$$

The optimum point is

$$\begin{bmatrix} x_1^* \\ x_2^* \\ x_3^* \\ \lambda^* \end{bmatrix} = \begin{bmatrix} 1 \\ 1 \\ 1 \\ -2 \end{bmatrix}$$

Let's find the Hessian of the Lagrange function.

$$\nabla^2_{\mathbf{x},\mathbf{x}}\mathcal{L} = \begin{bmatrix} \nabla_{x_1x_1}\mathcal{L} & \nabla_{x_1x_2}\mathcal{L} & \nabla_{x_1x_3}\mathcal{L} \\ \nabla_{x_2x_1}\mathcal{L} & \nabla_{x_2x_2}\mathcal{L} & \nabla_{x_2x_3}\mathcal{L} \\ \nabla_{x_3x_1}\mathcal{L} & \nabla_{x_3x_2}\mathcal{L} & \nabla_{x_3x_3}\mathcal{L} \end{bmatrix} = \begin{bmatrix} 0 & 1 & 1 \\ 1 & 0 & 1 \\ 0 & 0 & 1 \end{bmatrix}$$

To check the definiteness of the Hessian, the eigenvalues of the Hessian are determined as follows:

$$|\nabla^2_{\mathbf{x},\mathbf{x}}\mathcal{L} - \mu I| = 0 \tag{3.4}$$

which leads to $\mu_1 = 2$, $\mu_2 = -1$, and $\mu_3 = -1$. Eigenvalues of Hessian are not strictly positive or negative; hence, the optimal point x^* is a saddle point.

3.2 Inequality Constraints (KKT Conditions)

Inequality constraints are often found in function optimization. An optimization subject to inequality constraints is defined as follows:

$$\min_{\mathbf{x}} f(\mathbf{x})$$

subject to

$$h_j(\mathbf{x}) \leq 0;\ j = 1, 2...p$$

where $h_j(\mathbf{x}) = 0$ are the inequality constraints to be satisfied. To illustrate the situation, let's take the scalar function $f(x)$ minimization subject to the inequality constraint $h(x) \leq 0$ as shown in Fig. 3.2. As evident in the left figure, the global minima x^M is not feasible due to the positioning of the inequality constraint. At

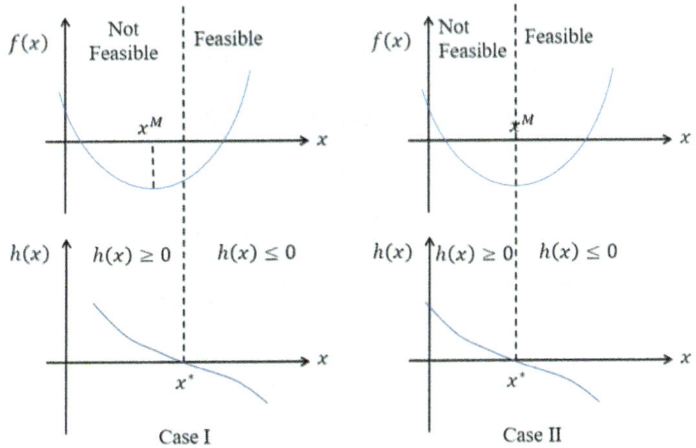

Fig. 3.2 Scalar function $f(x)$ minimization subject to inequality constraint $h(x) \leq 0$

the feasible minima x^* the function gradient is $\nabla f(x^*) > 0$ and the inequality constraint gradient is $\nabla h(x^*) < 0$. The two gradients have opposite signs. The same result will be reached when the inequality is moved to the left of x^M, in which case the polarity of the inequality has to change to make x^M infeasible because otherwise, the global minima falls within the inequality constraint and prevails. Extending to a general vector case, the first-order necessary condition for optimality subject to the inequality constraint is as follows:

$$\begin{aligned} \nabla_{\mathbf{x}} f(\mathbf{x}) + \mu \nabla_{\mathbf{x}} h(\mathbf{x}) &= \mathbf{0} \\ h(\mathbf{x}) \leq 0,\ \mu &> 0 \end{aligned} \tag{3.5}$$

However, there is another specific case, as shown in the right figure, where the inequality boundary $h(x) = 0$ coincides with the global minima x^*, in which case the first-order optimality condition is as follows:

$$\begin{aligned} \nabla_{\mathbf{x}} f(\mathbf{x}) + \mu \nabla_{\mathbf{x}} h(\mathbf{x}) &= \mathbf{0} \\ \mu h(\mathbf{x}) &= 0 \end{aligned} \tag{3.6}$$

Combining the two possibilities (3.5) and (3.6), the first-order condition for optimality under inequality constraints is defined as follows:

$$\begin{gathered} \nabla_{\mathbf{x}} f(\mathbf{x}) + \mu \nabla_{\mathbf{x}} h(\mathbf{x}) = \mathbf{0} \\ h(\mathbf{x}) \leq 0,\ \mu > 0 \ \text{ or } \ \mu h(\mathbf{x}) = 0 \end{gathered} \tag{3.7}$$

When both equality and inequality constraints are present, the first-order necessary condition for optimality subject to the m number of linear equality constraints and the p number of linear inequality constraints is constructed by combining (3.1) and (3.7) as follows:

$$\begin{aligned} \nabla_{\mathbf{x}} f(\mathbf{x}) + \boldsymbol{\lambda}^T \nabla_{\mathbf{x}} \mathbf{g}(\mathbf{x}) + \boldsymbol{\mu}^T \nabla_{\mathbf{x}} \mathbf{h}(\mathbf{x}) = 0 \\ \mathbf{g}(\mathbf{x}) = 0 \text{ and } h_j(\mathbf{x}) \le 0,\ \mu_j > 0 \text{ or } \mu_j h_j(\mathbf{x}) = 0 \end{aligned} \tag{3.8}$$

where $\mathbf{h}(\mathbf{x}) = [h_1(\mathbf{x})\ h_2(\mathbf{x})...h_m(\mathbf{x})]^T$ are m linear equality conditions and $\mathbf{g}(\mathbf{x}) = [g_1(\mathbf{x})\ g_2(\mathbf{x})...g_p(\mathbf{x})]^T$ are p linear inequality constraints. The condition (3.8) is also known as KKT (Karush–Kuhn–Tucker) conditions. The KKT conditions can be written in the following summation form as well.

$$\begin{aligned} \nabla_{\mathbf{x}} f(\mathbf{x}) + \sum_{i=1}^{m} \lambda_i \nabla_{\mathbf{x}} g_i(\mathbf{x}) + \sum_{j=1}^{p} \mu_j \nabla_{\mathbf{x}} h_j(\mathbf{x}) = 0 \\ \mathbf{g}(\mathbf{x}) = 0 \text{ and } g_i(\mathbf{x}) \le 0,\ \mu_j > 0 \text{ or } \mu_j h_j(\mathbf{x}) = 0 \end{aligned} \tag{3.9}$$

The constrained optimal solution produced by (3.9) might violate the KKT conditions. In such situations, some inequality constraints can be disregarded when (3.9) is applied. But the resulting optimal solution needs to be tested against all of the KKT conditions for validity. The following example illustrates how the KKT conditions are used.

Exercise: KKT Conditions

Find the optimality of the function

$$f(x_1, x_2) = x_1^2 + x_2^2 + 3x_1 - 4x_2$$

subject to the constraints

$$\begin{aligned} g_1(x_1, x_2) &= x_1 + 2x_2 \le 0 \qquad \text{and} \\ g_2(x_1, x_2) &= 2x_1 - x_2 - 1 \le 0 \end{aligned}$$

Solution:

Let's define the Lagrangian with the inequality constraints as follows:

$$\begin{aligned} L(x_1, x_2, \mu_1, \mu_2) = x_1^2 + x_2^2 + 3x_1 - 4x_2 + \mu_1(x_1 + 2x_2) + \\ \mu_2(2x_1 - x_2 - 1) \end{aligned}$$

Then, the FONCs are as follows:

$$\begin{aligned}
\nabla_{x_1} L &= 2x_1 + 3 + \mu_1 + 2\mu_2 = 0 \\
\nabla_{x_2} L &= 2x_2 - 4 + 2\mu_1 - \mu_2 = 0 \\
\nabla_{\mu_1} L &= x_1 + 2x_2 = 0 \\
\nabla_{\mu_2} L &= 2x_1 - x_2 - 1 = 0
\end{aligned}$$

The solution of the four equations above is $\mathbf{x}^* = [1, -2]^T$ and $\boldsymbol{\mu}^* = [11/5, -18/5]^T$. However, this solution violates KKT condition $\mu_i > 0$ in (3.9). Therefore, the above solution is not acceptable.

Relaxation of the Constraints

Let's check if the unconstrained optimality is acceptable under the inequality constraints. For unconstrained optimization, the first-order condition for optimality is as follows:

$$\begin{aligned}
\nabla_{x_1} f &= 2x_1 + 3 = 0 \\
\nabla_{x_2} f &= 2x_2 - 4 = 0
\end{aligned}$$

The solution is $\mathbf{x}^* = [-\frac{3}{2}\ 2]^T$. However, this solution also violates the KKT conditions as $g_1(\mathbf{x}^*) = 2.5 \nleq 0$.

Introduce Only the First Constraint

Let's introduce only the first inequality constraint; for that, the Lagrangian is defined as follows:

$$L(x_1, x_2, \mu_1) = x_1^2 + x_2^2 + 3x_1 - 4x_2 + \mu_1(x_1 + 2x_2)$$

Then the FONCs are as follows:

$$\begin{aligned}
\nabla_{x_1} L &= 2x_1 + 3 + \mu_1 = 0 \\
\nabla_{x_2} L &= 2x_2 - 4 + 2\mu_1 = 0 \\
\nabla_{\mu_1} L &= x_1 + 2x_2 = 0
\end{aligned}$$

And the solution is $\mathbf{x}^* = [-2, 1]^T$ and $\mu_1^* = 1$. This solution satisfies the KKT conditions because it satisfies the second inequality constraint $g_2(\mathbf{x}^*) = -6 \leq 0$ as well. Moreover, the solution indicates the fact that $g_1(\mathbf{x}^*) = 0$

3.3 Constrained Optimization in Matlab

3.3.1 Quadratic Optimization with Linear Constraints

The Matlab `quadprog` command implements quadratic function minimization subject to linear constraints. The command is launched with the arguments as `>> quadprog`($\mathbf{Q}, \mathbf{r}, \mathbf{A}, \mathbf{b}, \mathbf{A_{eq}}, \mathbf{b_{eq}}, \mathbf{LB}, \mathbf{UB}$) where $\mathbf{Q}, \mathbf{r}$ are quadratic cost function parameters given in (2.7), $\mathbf{A}, \mathbf{b}$ define the linear inequality constraints, $\mathbf{A_{eq}}, \mathbf{b_{eq}}$ define the linear equality constraints, and **LB**, **UB** define the lower and upper bounds of the variables.

Example: Linear Inequality Constraints

Find the minima of the following quadratic function:

$$f(x_1, x_2) = 0.4x_1^2 + x_2^2 - 5x_1 - 6x_2 + 50$$

subject to the following inequality constraints:

$$\begin{aligned} x_1 - x_2 &\leq -2 \\ -0.3x_1 - x_2 &\leq -8, \end{aligned}$$

and within the following bounds of the variables:

$$\begin{aligned} 0 \leq x_1 \leq 10 \\ 1 \leq x_2 \leq 8 \end{aligned}$$

From the cost function, the $\mathbf{Q}$ and $\mathbf{r}$ are identified as follows:

$$\begin{aligned} f(x_1, x_2) &= 0.4x_1^2 + x_2^2 - 5x_1 - 6x_2 + 50 \\ &= \frac{1}{2}\begin{bmatrix} x_1 & x_2 \end{bmatrix}\begin{bmatrix} 0.8 & 0 \\ 0 & 2 \end{bmatrix}\begin{bmatrix} x_1 \\ x_2 \end{bmatrix} + \begin{bmatrix} -5 & -6 \end{bmatrix}\begin{bmatrix} x_1 \\ x_2 \end{bmatrix} + 50 \\ &= \frac{1}{2}\mathbf{x}^T\mathbf{Q}\mathbf{x} + \mathbf{r}^T\mathbf{x} + c \end{aligned}$$

Therefore,

$$\mathbf{Q} = \begin{bmatrix} 0.8 & 0 \\ 0 & 2 \end{bmatrix}, \quad \mathbf{r} = \begin{bmatrix} -5 \\ -6 \end{bmatrix}$$

The linear inequalities are arranged as follows:

$$\begin{bmatrix} 1 & -1 \\ -0.3 & -1 \end{bmatrix}\begin{bmatrix} x_1 \\ x_2 \end{bmatrix} \leq \begin{bmatrix} -2 \\ -8 \end{bmatrix}$$

Therefore,

$$\mathbf{A} = \begin{bmatrix} 1 & -1 \\ -0.3 & -1 \end{bmatrix}, \quad \mathbf{b} = \begin{bmatrix} -2 \\ -8 \end{bmatrix}$$

There are no linear equality constraints; therefore,

$$\mathbf{A}_{eq} = [\,], \quad \mathbf{b}_{eq} = [\,]$$

The lower and upper bounds of (x_1, x_2) are specified by the following vectors:

$$\mathbf{LB} = \begin{bmatrix} 0 \\ 1 \end{bmatrix}, \quad \mathbf{UB} = \begin{bmatrix} 10 \\ 8 \end{bmatrix}$$

The Matlab code given in Appendix C uses the `quadprog` command to carry out the constrained quadratic optimization and produces the result shown in Fig. 3.3. The `quadprog` determines the optimum point $\mathbf{x}^* = [4.6\ 6.6]^T$ and the constrained minimum value $f(\mathbf{x}^*) = -10.5 + 50 = 39.5$.

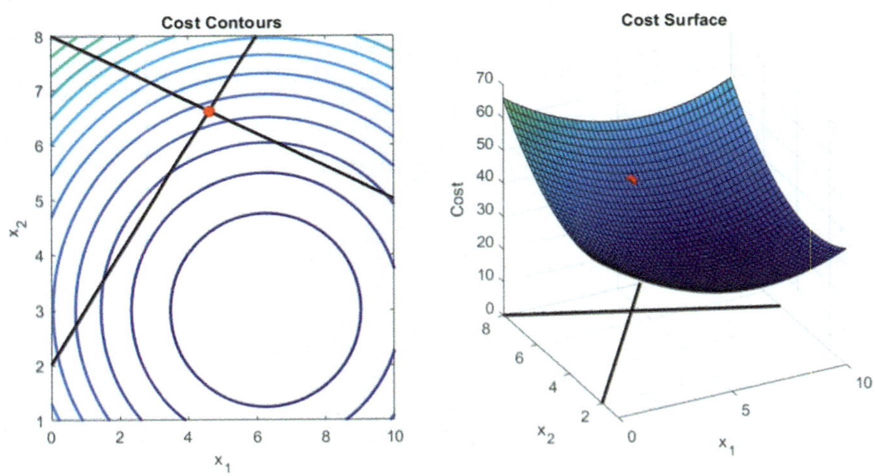

Fig. 3.3 Cost contour lines, cost surface, two linear inequality constraints, and the feasible minimum cost

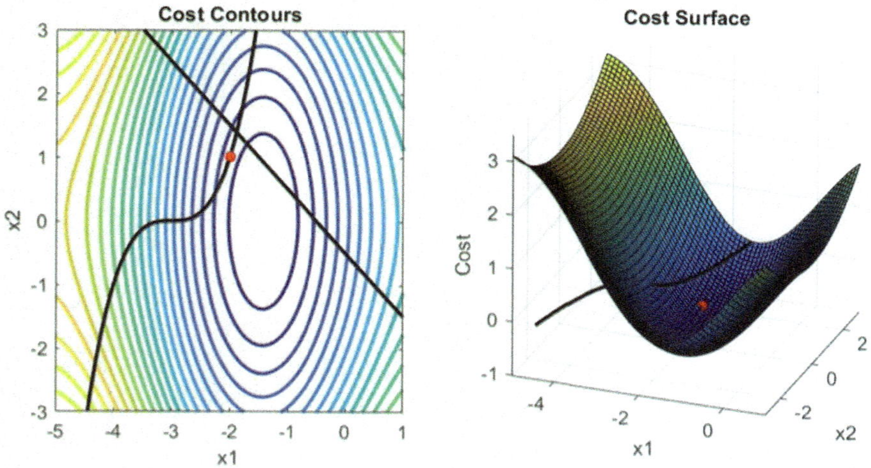

Fig. 3.4 Cost contour lines, cost surface, linear and nonlinear inequality constraints, and the feasible minimum cost

3.3.2 Optimization of Arbitrary Functions Subject to Nonlinear Constraints

The Matlab function `fmincon` implements arbitrary function optimization subject to nonlinear constraints. The optimization command is launched as follows: >> `fmincon(@(x)cost(x), x0, A, b, Aeq, Beq, LB, UB, @(x)NLcon(x), options)`. The `cost(x)` is any arbitrary function to be minimized. Arguments `A`, `b`, $\mathtt{A_{eq}}$, $\mathtt{b_{eq}}$, `LB`, and `UB` have already been described in the `quadprog` section. The optimization process starts from the initial point `x(0)`.

Example: Nonlinear Equality Constraints

Find the minimum of $f(\mathbf{x}) = \sin(x_1) + 0.05x_1^2 + 0.1x_2^2$ subject to the linear inequality constraint described by $A = [1\ 1]$; $B = [-0.5]$ and the nonlinear equality constraint $(x_1 + 3)^3 - x_2 = 0$. The Matlab code in Appendix D carries out the optimization process, and the results are shown in Fig. 3.4. The optimum point is $\mathbf{x}^* = [-2\ 1]$ and the minimum feasible value is $f(\mathbf{x}^*) = -0.61$

The next chapter presents unconstrained functional optimization, in which the optimum function is selected from a group of functions without considering the constraints. The implementation in Matlab is also demonstrated.

Chapter 4
Unconstrained Functional Optimization

This chapter presents how to use the calculus of variations to find the optimum function within a group of functions without considering the constraints. A few examples of functional optimization, including the shortest path and the minimum-time path, are presented. The Matlab implementation of functional optimization is also demonstrated.

4.1 Calculus of Variations

Figure 4.1 shows possible paths $f(x)$ of a single variable functional between two fixed points P_1 and P_2. It also shows the optimal function $f^*(x)$.

The problem to be solved in functional optimization is to determine $f^*(\mathbf{x})$ using the calculus of variations. A few classical problems of functional optimization are presented below.

Shortest Path

Figure 4.2 shows possible paths between two points P_1 and P_2 on the 2D plane. Determine the shortest path between these two points.

The path length from x_1 to x_2 is determined as follows:

S. R. Munasinghe, *Optimization and Optimal Control in a Nutshell*, Engineering Optimization: Methods and Applications,
https://doi.org/10.1007/978-981-97-8167-6_4

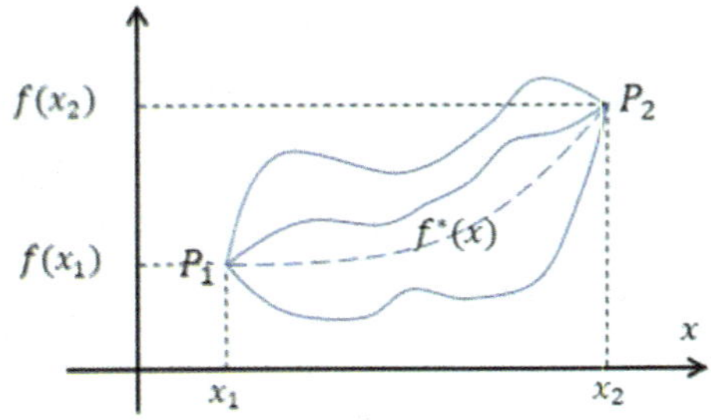

Fig. 4.1 Set of possible paths (functional) and the optimal path between two points

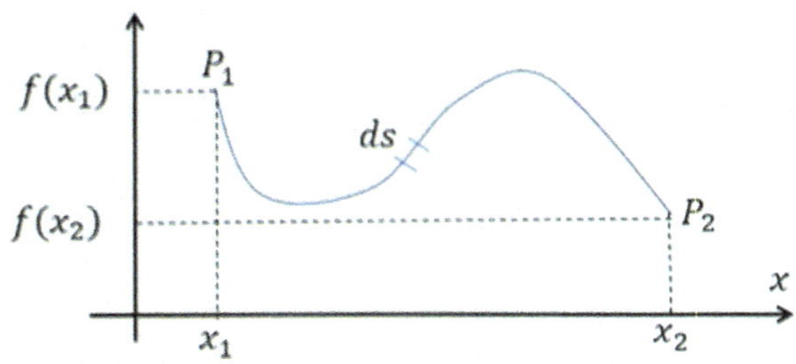

Fig. 4.2 Finding the shortest path between two points on a plane

$$
\begin{aligned}
L(x, f(x)) &= \int_{x_1}^{x_2} ds \\
&= \int_{x_1}^{x_2} \sqrt{dx^2 + df(x)^2} \\
&= \int_{x_1}^{x_2} \sqrt{1 + \left(\frac{df(x)}{dx}\right)^2}\, dx \\
&= \int_{x_1}^{x_2} \sqrt{1 + \dot{f}^2(x)dx}
\end{aligned}
\tag{4.1}
$$

Then, the optimization problem is to determine the particular function $f^*(x)$ that minimizes this integral. We will solve this at a later stage.

Minimum-Time Path

Figure 4.3 shows a possible path of a particle moving from P_1 to P_2 with a variable speed. Determine the minimum-time path between the two points.

The time taken to move from P_1 to P_2 is determined as follows:

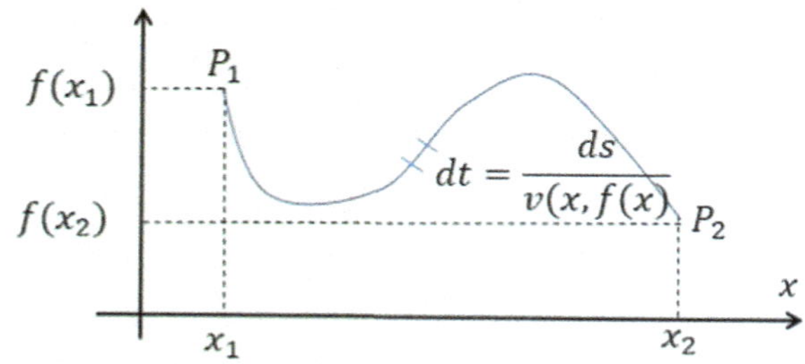

Fig. 4.3 Finding the minimum-time path between two points

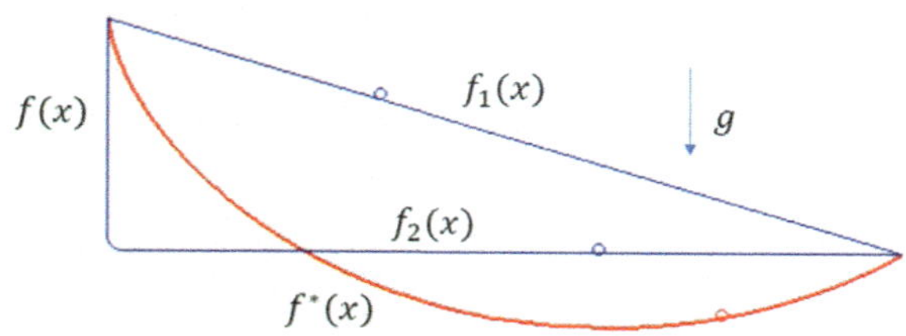

Fig. 4.4 The Brachistocrone is the minimum-time sliding path of a mass falling under gravity. Image source: Wikipedia

$$
\begin{aligned}
T(x, f(x)) &= \int_{x_1}^{x_2} dt \\
&= \int_{x_1}^{x_2} \frac{\sqrt{1 + \left(\frac{df(x)}{dx}\right)^2}}{v(x, f(x))} dx \\
&= \int_{x_1}^{x_2} \frac{\sqrt{1 + \dot{f}^2(x)}}{v(x, f(x))} dx
\end{aligned}
$$

The optimization problem is to determine the particular function $f^*(x)$ that minimizes this integral. This is the well-known "Brachistocrone" problem posed for solutions by Yohann Bernaulli in 1696. A number of solutions were proposed by several great mathematicians, including Sir Isaac Newton, but the direct solution was eventually formulated by Yohann Bernauli himself in 1697. It was found that the minimum-time path $f^*(x)$ is a cycloid, as shown below (Fig. 4.4).

4.2 The Conditions for Functional Optimization

Based on the two examples above, the general condition for optimization of a single variable functional $f(x)$ can be formulated as follows: Find the function $f^*(x) \in f(x)$ that optimizes the integral

$$
J(f(x)) = \int_{x_1}^{x_2} F(x, f(x), \dot{f}(x)) dx \tag{4.2}
$$

Let's create the functional $f(x)$ illustrated in Fig. 4.1 using arbitrary perturbation $\epsilon\eta(x)$ from the optimal function $f^*(x)$ as follows:

$$f(x) = f^*(x) + \epsilon\eta(x) \tag{4.3}$$

where $\eta(x)$ is an arbitrary function with zero value at the boundary points $\eta(x_1) = \eta(x_2) = 0$ as shown in Fig. 4.1, and ϵ is a random variable used to create an infinite number of functions from the arbitrary perturbation function $\eta(x)$. The optimal function $f^*(x)$ is referred to as $\epsilon = 0$. The functional integration in (4.2) can be written with respect to ϵ, as shown below, because the functional $f(x)$ is spanned by ϵ.

$$J(\epsilon) = \int_{x_1}^{x_2} F(x, f(x), \dot{f}(x))dx \tag{4.4}$$

Then, the condition leading to the optimal function $f^*(x)$ is as follows:

$$\begin{aligned}
\frac{d}{d\epsilon}\int_{x_1}^{x_2} F(x, f(x), \dot{f}(x))dx = 0 \\
\int_{x_1}^{x_2} \frac{\partial}{\partial\epsilon} F(x, f(x), \dot{f}(x))dx = 0 \\
\int_{x_1}^{x_2} \left[\frac{\partial F}{\partial f}\frac{\partial f}{\partial\epsilon} + \frac{\partial F}{\partial \dot{f}}\frac{\partial \dot{f}}{\partial\epsilon}\right] dx = 0
\end{aligned}$$

From (4.3), $\frac{\partial f}{\partial\epsilon} = \eta$ and $\frac{\partial \dot{f}}{\partial\epsilon} = \dot{\eta}$, therefore

$$\int_{x_1}^{x_2} \left[\frac{\partial F}{\partial f}\eta + \frac{\partial F}{\partial \dot{f}}\dot{\eta}\right] dx = 0 \tag{4.5}$$

When integrated by parts ($\int u\dot{v} = uv| - \int v du$) with $u = \frac{\partial F}{\partial \dot{f}}$ and $v = \eta$ the second term of (4.5) can be written as follows:

$$\begin{aligned}
\int_{x_1}^{x_2} \frac{\partial F}{\partial \dot{f}}\dot{\eta}dx &= \frac{\partial F}{\partial \dot{f}}\eta|_{x_1}^{x_2} - \int_{x_1}^{x_2} \eta\frac{d}{dx}\left(\frac{\partial F}{\partial \dot{f}}\right)dx \\
&= \frac{\partial F}{\partial \dot{f}}[\eta(x_2) - \eta(x_1)] - \int_{x_1}^{x_2} \eta\frac{d}{dx}\left(\frac{\partial F}{\partial \dot{f}}\right)dx \\
&= -\int_{x_1}^{x_2} \eta\frac{d}{dx}\left(\frac{\partial F}{\partial \dot{f}}\right)dx
\end{aligned} \tag{4.6}$$

Because $\eta(x_1) = \eta(x_2) = 0$. When (4.6) is substituted in (4.5) the following result is obtained:

$$\int_{x_1}^{x_2} \left[\frac{\partial F}{\partial f}\eta - \eta \frac{d}{dx}\left(\frac{\partial F}{\partial \dot{f}}\right) \right] dx = 0$$

$$\int_{x_1}^{x_2} \left[\frac{\partial F}{\partial f} - \frac{d}{dx}\left(\frac{\partial F}{\partial \dot{f}}\right) \right] \eta dx = 0 \tag{4.7}$$

Because $\eta \neq 0$ is an arbitrary function, the condition for (4.7) is as follows:

$$\frac{\partial F}{\partial f} - \frac{d}{dx}\left(\frac{\partial F}{\partial \dot{f}}\right) = 0 \tag{4.8}$$

This is the Lagrange–Euler equation for functional optimality. When this condition is satisfied, $\epsilon \to 0$, and $f(x) \to f^*(x)$.

Example: Geodesic on a Plane

Geodesic is the shortest path between two points. To find the shortest path function, the functional optimization (4.2) is applied to the path length function of (4.1). Hence, the integrand of the functional optimization is $F = \sqrt{1 + \dot{f}^2(x)}$. Then, the following statements can be derived:

$$\frac{\partial F}{\partial f} = 0 \tag{4.9}$$

$$\frac{\partial F}{\partial \dot{f}(x)} = \frac{\dot{f}(x)}{\sqrt{1 + \dot{f}^2(x)}} \tag{4.10}$$

Then the Euler–Lagrange condition for functional optimization (4.8) is used as follows:

$$0 - \frac{d}{dx}\left(\frac{\dot{f}(x)}{\sqrt{1 + \dot{f}^2(x)}}\right) = 0$$

which leads to the following:

$$-\frac{\dot{f}(x)}{\sqrt{1 + \dot{f}^2(x)}} = c$$

$$\dot{f}(x) = \frac{c}{\sqrt{1 - c^2}} = c_1$$

$$f(x) = c_1 x + c_2 \tag{4.11}$$

where c, c_1, c_2 are constants. The result (4.11) indicates that the shortest path between two points on a plane (geodesic) is a straight line, which is a known fact, and the result therefore conforms to the application of the Euler–Lagrange equation in determining the optimal function in a functional.

Exercise: Geodesic on a Sphere

Find the shortest path between two points on a sphere.

4.3 Beltrami Identity

When the integrand function does not depend on the independent variable x it takes the form of $F(f(x), \dot{f}(x))$. Then, the Beltrami identity provides the first-order necessary condition for optimality. When F is independent of x.

$$\begin{aligned}\frac{dF}{dx} &= \frac{\partial F}{\partial f}\dot{f} + \frac{\partial F}{\partial \dot{f}}\ddot{f} = 0 \\ \frac{\partial F}{\partial f}\dot{f} &= -\frac{\partial F}{\partial x} - \frac{\partial F}{\partial \dot{f}}\ddot{f}\end{aligned} \tag{4.12}$$

Multiplying the Euler–Lagrange equation (4.8) by $\dot{f}$

$$\frac{\partial F}{\partial f}\dot{f} = \frac{d}{dx}\left(\frac{\partial F}{\partial \dot{f}}\right)\dot{f} \tag{4.13}$$

From (4.12) and (4.13),

$$\begin{aligned}-\frac{\partial F}{\partial x} - \frac{\partial F}{\partial \dot{f}}\ddot{f} &= \frac{d}{dx}\left(\frac{\partial F}{\partial \dot{f}}\right)\dot{f} \\ \frac{\partial F}{\partial x} + \frac{\partial F}{\partial \dot{f}}\ddot{f} + \frac{d}{dx}\left(\frac{\partial F}{\partial \dot{f}}\right)\dot{f} &= 0 \\ \frac{\partial F}{\partial x} + \frac{d}{dx}\left(\dot{f}\frac{\partial F}{\partial \dot{f}}\right) &= 0 \\ \frac{d}{dx}\left[F + \dot{f}\frac{\partial F}{\partial \dot{f}}\right] &= 0 \\ \text{Therefore,}\quad F + \dot{f}\frac{\partial F}{\partial \dot{f}} &= c\end{aligned} \tag{4.14}$$

The Beltrami identity is used in place of the Euler–Lagrange equation when the function F does not depend on the independent variable.

4.4 Functional Optimization in Matlab

The following examples illustrate how functional optimization is implemented in Matlab (https://www.mathworks.com/help/symbolic/sym.functionalderivative.html).

Example 1

Find the function $y(x)$ that optimizes the objective function $J(y(x)) = \int_a^b y(x)\sin(y(x))dx$.

Solution:

The following symbolic Matlab code

```
clear; close all;
syms y(x)        % declare symbolic function y(x)
F = y*sin(y);    % declare cost function
G = functionalDerivative(F,y)
```

determines the variational derivative as $\frac{dJ(y(x))}{dy(x)} = \sin(y(x)) + \cos(y(x))y(x)$. The optimal function is the one that has zero functional derivative; hence, the optimal function $y(x)$ is

$$\sin(y(x)) + \cos(y(x))y(x) = 0.$$

Example 2

Find the equation of motion of the mass–spring system shown in Fig. 4.5 using the Lagrange–Euler formulation.

Solution:

The Lagrange–Euler Formulation needs the kinetic and potential energies of the system. The kinetic energy of the system is $K = \frac{1}{2}m\dot{x}^2(t)$ and the potential energy in the spring is $P = \int_0^x kx(t)dx(t) = \frac{1}{2}kx^2(t)$. The motion of the object is associated with the difference between P and K. The natural motion is the optimal path of the object, which corresponds to the minimum action associated with the Lagrangian $L = P - K$. The following Matlab code

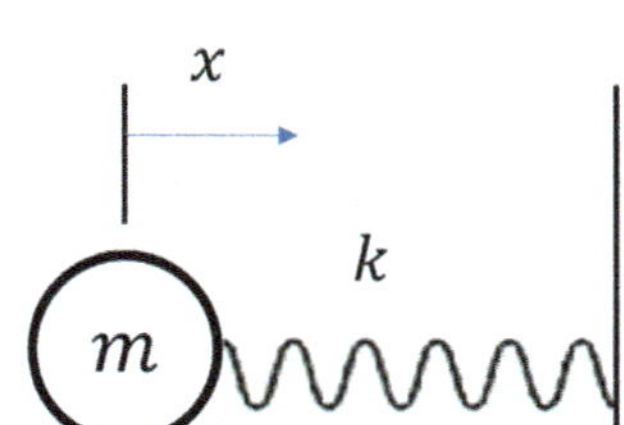

Fig. 4.5 Frictionless mass–spring system

```
clear; close all;
syms m k x(t)          % Declare m k x symbolic variables
K = 1/2*m*diff(x,t)^2;  % Declare kinetic energy
P = 1/2*k*x^2;          % Declare potential energy
L = K - P;              % declare Lagrangian function
functionalDerivative(L,x)
```

generates the variational derivative of the Lagrangian as $m\ddot{x}(t) + kx(t)$. The optimum path $x(t)$ is the one that has zero functional derivatives; hence, the optimum path is determined as follows:

$$m\ddot{x}(t) + kx(t) = 0$$

The next chapter presents a constrained functional optimization process with several examples.

Chapter 5
Constrained Functional Optimization

This chapter presents how to introduce constraints into the functional optimization process. Two examples of constrained functional optimization—the catenary problem and the isoperimetric problem—are presented.

5.1 The Augmented Lagrangian

When constraints exist, the functional optimization problem (4.2) is defined as follows: Find the optimal function f^* that optimizes the following integral

$$J = \int_{x_1}^{x_2} L(x, f(x), \dot{f}(x))dx \tag{5.1}$$

subject to the constraints

$$\mathbf{g} = g_i(x, f(x), \dot{f}(x)) = 0; i = 1, 2 \ldots p \tag{5.2}$$

Following the process of including constraints in the optimization process as shown in (3.2), an augmented Lagrangian L^a is defined as follows:

$$L^a = L + \boldsymbol{\lambda}^T \mathbf{g} \tag{5.3}$$

Then the integral in (5.1) changes as follows:

$$J = \int_{x_1}^{x_2} L^a(x, f(x), \dot{f}(x), \boldsymbol{\lambda})dx \tag{5.4}$$

S. R. Munasinghe, *Optimization and Optimal Control in a Nutshell*, Engineering Optimization: Methods and Applications,
https://doi.org/10.1007/978-981-97-8167-6_5

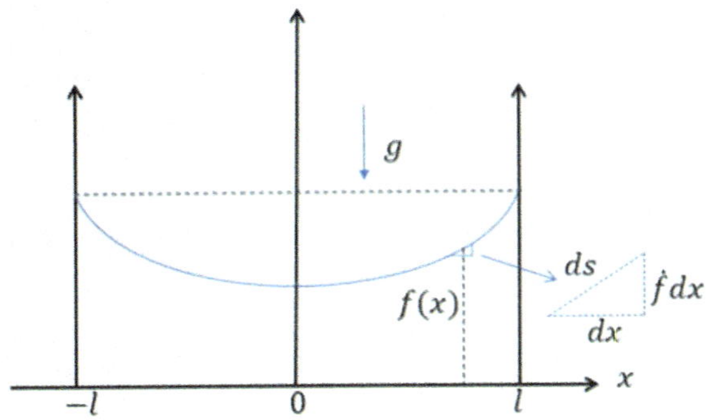

Fig. 5.1 Catenary: the shape of the string knotted to two poles at the same height

where $\boldsymbol{\lambda}^T = [\lambda_1, \lambda_2 ... \lambda_p]$ are Lagrange multipliers associated with the constraints. When the first-order necessary condition for optimality (4.8) is applied to (5.3),

$$\frac{\partial}{\partial f}(L + \boldsymbol{\lambda}^T \mathbf{g}) - \frac{d}{dx}\left[\frac{\partial}{\partial \dot{f}}(L + \boldsymbol{\lambda}^T \mathbf{g})\right] = 0$$
$$\frac{\partial L}{\partial f} - \frac{d}{dx}\left(\frac{\partial L}{\partial \dot{f}}\right) + \boldsymbol{\lambda}^T\left[\frac{\partial \mathbf{g}}{\partial f} - \frac{d}{dx}\left(\frac{\partial \mathbf{g}}{\partial \dot{f}}\right)\right] = 0 \quad (5.5)$$

Several examples where augmented Lagrangian is used are presented below.

The Catenary Problem

Figure 5.1 shows the catenary, the shape of the string knotted onto two poles at the same height. To find the shape of the catenary, the potential energy of the string is minimized, subject to the string length constraint.

The total potential energy of the string of length S is

$$L = \int_0^S \rho.ds.g.f(x)$$
$$= \int_{-l}^{l} \rho g \sqrt{1 + \dot{f}^2(x)} f(x) dx \quad (5.6)$$

where ρ is the linear density of the string and $ds = \sqrt{1 + \dot{f}^2(x)}dx$. The constraint is the length S of the string.

$$S = \int_{-l}^{l} \sqrt{1 + \dot{f}^2(x)} dx$$

Therefore, the string length constraint is written as follows:

$$g = \int_{-l}^{l} \sqrt{1 + \dot{f}^2(x)} dx - S = 0 \quad (5.7)$$

Then, the optimization problem is stated as follows: Find $f^*(x)$ that minimizes the total potential energy of the string (5.6) subject to the total length constraint (5.7). Combining these two equations, let's define the augmented Lagrangian with a single Lagrange multiplier λ as follows:

$$\begin{aligned} L^a &= \int_{-l}^{l} \rho g \sqrt{1+\dot{f}^2(x)}\, f(x) + \lambda \left(\sqrt{1+\dot{f}^2(x)} dx - S \right) dx \\ &= \int_{-l}^{l} F(f(x), \dot{f}(x)) dx \end{aligned} \tag{5.8}$$

This Lagrangian does not depend on the independent variable x; hence, the Beltrami identity (4.14) can be used as the FONC for optimality, which will reveal the shape of the catenary $f^*(x)$. The reader is directed to Gander and Oswald (2004) for the rest of the solution.

The Isoperimetric Problem

Referring to Fig. 5.2, find the shape of the wire that maximizes the area under it, subject to the length of the wire S.

The area under the curve is

$$L = \int_{x_a}^{x_b} f(x) dx \tag{5.9}$$

And the length of the wire is

$$S = \int_{x_a}^{x_b} \sqrt{1+\dot{f}^2(x)} dx \tag{5.10}$$

Then, the constraint optimization problem is to find the function $f^*(x)$ that maximizes L subject to the constraint g. Let's define the augmented Lagrangian as follows:

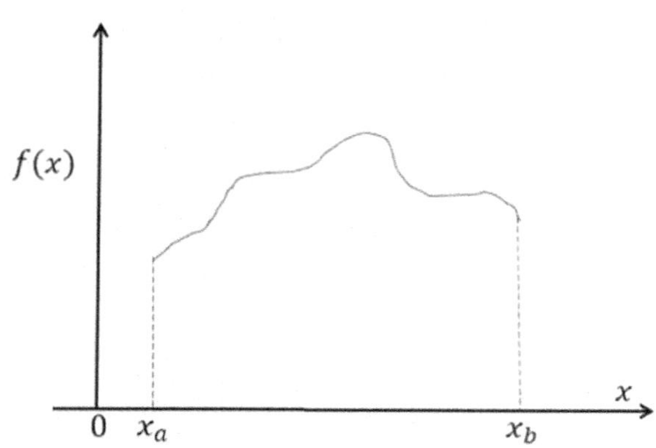

Fig. 5.2 Isoperimetric problem: Maximize the area under the curve

$$
\begin{aligned}
L^a &= \int_{x_a}^{x_b} \left[f(x) + \lambda\sqrt{1+\dot{f}^2(x)} - S \right] dx \\
&= \int_{x_a}^{x_b} F(f(x), \dot{f}(x)) dx
\end{aligned}
\tag{5.11}
$$

The optimum shape of the wire $f^*(x)$ can be obtained by applying the Euler–Lagrange equation (4.8) to the function F. Alternatively, the Beltrami identity (4.14) can be used in place of the Lagrange–Euler equation because F does not depend on x. The rest of the solution is available in Nahin (2003).

5.2 The Nature of Optimal Function

From the functional (4.3) and functional integral (4.4) the nature of the optimal function $f^*(x)$ is determined by the Hessian as follows: The abbreviated symbols f and $\dot{f}$ are used instead of $f(x)$ and $\dot{f}(x)$ for simplicity.

$$
\begin{aligned}
\left.\frac{d^2 J(\epsilon)}{d\epsilon^2}\right|_{\epsilon=0} &= \frac{d^2}{d\epsilon^2}\left[\int_{x_1}^{x_2} F(x, f, \dot{f}) dx\right]_{\epsilon=0} \\
&= \left.\int_{x_1}^{x_2} \frac{d^2}{d\epsilon^2} F(x, f, \dot{f}) dx\right|_{\epsilon=0} \\
&= \int_{x_1}^{x_2} \frac{\partial}{\partial\epsilon}\left[\frac{\partial}{\partial\epsilon} F(x, f, \dot{f}) dx\right]_{\epsilon=0} \\
&= \left.\int_{x_1}^{x_2} \frac{\partial}{\partial\epsilon}\left[\frac{\partial F}{\partial f}\frac{\partial f}{\partial\epsilon} + \frac{\partial F}{\partial \dot{f}}\frac{\partial \dot{f}}{\partial\epsilon}\right] dx\right|_{\epsilon=0}
\end{aligned}
$$

From (4.3), $\frac{\partial f}{\partial\epsilon} = \eta$ and $\frac{\partial \dot{f}}{\partial\epsilon} = \dot{\eta}$, therefore

$$
\begin{aligned}
\left.\frac{d^2 J(\epsilon)}{d\epsilon^2}\right|_{\epsilon=0} &= \left.\int_{x_1}^{x_2} \frac{\partial}{\partial\epsilon}\left[\frac{\partial F}{\partial f}\eta + \frac{\partial F}{\partial \dot{f}}\dot{\eta}\right] dx\right|_{\epsilon=0} \\
&= \left.\int_{x_1}^{x_2} \left[\eta\left(\frac{\partial^2 F}{\partial f^2}\frac{\partial f}{\partial\epsilon} + \frac{\partial^2 F}{\partial \dot{f}\partial f}\frac{\partial \dot{f}}{\partial\epsilon}\right) + \dot{\eta}\left(\frac{\partial^2 F}{\partial f\partial \dot{f}}\frac{\partial f}{\partial\epsilon} + \frac{\partial^2 F}{\partial \dot{f}\partial \dot{f}}\frac{\partial \dot{f}}{\partial\epsilon}\right)\right] dx\right|_{\epsilon=0} \\
&= \left.\int_{x_1}^{x_2} \left[\eta\left(\frac{\partial^2 F}{\partial f^2}\eta + \frac{\partial^2 F}{\partial \dot{f}\partial f}\dot{\eta}\right) + \dot{\eta}\left(\frac{\partial^2 F}{\partial f\partial \dot{f}}\eta + \frac{\partial^2 F}{\partial \dot{f}^2}\dot{\eta}\right)\right] dx\right|_{\epsilon=0} \\
&= \left.\int_{x_1}^{x_2} \left[\eta^2\frac{\partial^2 F}{\partial f^2} + 2\eta\dot{\eta}\frac{\partial^2 F}{\partial f\partial \dot{f}} + \dot{\eta}^2\frac{\partial^2 F}{\partial \dot{f}^2}\right] dx\right|_{\epsilon=0}
\end{aligned}
\tag{5.12}
$$

If the cost function $J(\epsilon)$ is quadratic and convex, then the optimality is a minima, as shown in Fig. 5.3.

The next chapter presents how Lagrange–Euler and Pontriyagin–Hamiltonian formulations are used in optimal control design for linear, continuous-time systems.

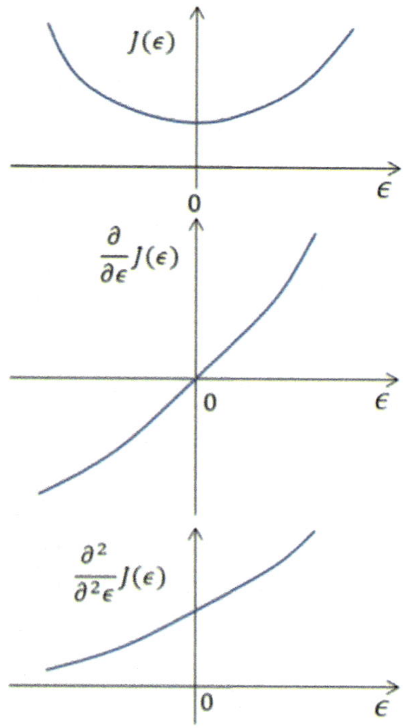

Fig. 5.3 Variation of performance index of a functional in which the functions are quadratic convex

References

Gander W, Oswald U (2004) The catenary curve. In: Solving problems in scientific computing using Maple and MATLAB®. Springer, Berlin, Heidelberg. https://doi.org/10.1007/978-3-642-18873-2_29

Nahin, PJ (2003) 6.8 The isoperimetric problem, solved (at last!). When least is best: how mathematicians discovered many clever ways to make things as small (or as large) as possible. Princeton University Press, Princeton, pp 251–259. https://doi.org/10.1515/9780691220383-046

Chapter 6
Continuous-Time Optimal Control

This chapter presents the optimal control design for linear, continuous-time systems. The two main formulations of functional optimization—Lagrange–Euler and Pontryagin–Hamiltonian—are presented, and the conditions for optimality in each formulation are derived. The nature of optimality is also explored. The process of introducing end-point conditions of final time and final state is presented. Several exercises are provided to emphasize the application of the two formulations in optimal control system design.

6.1 Lagrangian Formulation

Plant dynamics is often modeled using a set of state variables $\mathbf{x}(t) = [x_i(t)]; i = 1, 2 \ldots n$ where time t is the independent variable. Hence, the general condition for optimality (4.8) is re-written with $x \rightarrow t$ and $f(x) \rightarrow \mathbf{x}(t)$. In optimal control, there is a final state to achieve, and the error at the final state $\mathbf{x}_{t_f}$ is considered the terminal cost. Hence, the total cost of control from t_0 to t_f can be written as follows:

$$J = S(t_f, \mathbf{x}_{t_f}) + \int_{t_0}^{t_f} L(t, \mathbf{x}(t), \mathbf{u}(t)) + \boldsymbol{\lambda}^T(\mathbf{f}(t, \mathbf{x}(t), \mathbf{u}(t)) - \dot{\mathbf{x}})dt \tag{6.1}$$

where $S(t_f, \mathbf{x}_{t_f})$ is the terminal cost, which is zero if the system achieves the desired final state. The terminal cost can be written as the following integral:

$$S(t_f, \mathbf{x}_{t_f}) = \int_{t_0}^{t_f} \frac{d}{dt} S(t, \mathbf{x}(t))dt + S(t_0, \mathbf{x}_{t_0}) \tag{6.2}$$

S. R. Munasinghe, *Optimization and Optimal Control in a Nutshell*, Engineering Optimization: Methods and Applications,
https://doi.org/10.1007/978-981-97-8167-6_6

Let's set the final time t_f and final state $\mathbf{x}(t_f)$ free, giving enough time and freedom for the system to reach its own final state at its own final time. When substituted from (6.2) to (6.1)

$$J = \int_{t_0}^{t_f} \frac{d}{dt} S(t, \mathbf{x}(t))dt + S(t_0, \mathbf{x}_{t_0}) + \int_{t_0}^{t_f} L(t, \mathbf{x}(t), \mathbf{u}(t)) + \boldsymbol{\lambda}^T (\mathbf{f}(t, \mathbf{x}(t), \mathbf{u}(t)) - \dot{\mathbf{x}})dt \tag{6.3}$$

Let's write the first term using partial derivatives and drop the second term because it is a constant.

$$J = \int_{t_0}^{t_f} \left\{ \frac{\partial}{\partial \mathbf{x}} S(t, \mathbf{x}(t))\dot{\mathbf{x}}(t) + \frac{\partial}{\partial t} S(t, \mathbf{x}(t)) + L(t, \mathbf{x}(t), \mathbf{u}(t)) + \boldsymbol{\lambda}^T (\mathbf{f}(t, \mathbf{x}(t), \mathbf{u}(t)) - \dot{\mathbf{x}}) \right\} dt \tag{6.4}$$

$$= \int_{t_0}^{t_f} L^a(t, \mathbf{x}(t), \dot{\mathbf{x}}(t), \mathbf{u}(t), \boldsymbol{\lambda}(t))dt \tag{6.5}$$

where L^a is the augmented Lagrangian. Let's assume that the optimal cost corresponds to the optimal control $\mathbf{u}^*(t)$. And, a small deflection $\delta\mathbf{u}$ in control deflects the augmented Lagrangian to $L^{a\delta}$, final time by δt_f, and cost by δJ as follows:

$$\begin{aligned}
\delta J &= J(\mathbf{u}^*(t) + \delta\mathbf{u}) - J(\mathbf{u}^*(t)) \\
&= \int_{t_0}^{t_f+\delta t_f} L^{a\delta} dt - \int_{t_0}^{t_f} L^a dt \\
&= \int_{t_0}^{t_f} \left(L^{a\delta} - L^a\right) dt + \int_{t_f}^{t_f+\delta t_f} L^{a\delta} dt
\end{aligned} \tag{6.6}$$

Using the Taylor series approximation,

$$\begin{aligned}
\delta J &\approx \int_{t_0}^{t_f} \left\{ L^a + \left(\frac{\partial L^a}{\partial \mathbf{x}}\right)^T \delta\mathbf{x} + \left(\frac{\partial L^a}{\partial \dot{\mathbf{x}}}\right)^T \delta\dot{\mathbf{x}} + \left(\frac{\partial L^a}{\partial \mathbf{u}}\right)^T \delta\mathbf{u} - L^a \right\} dt + L^a|_{t_f} \delta t_f \\
&\approx \int_{t_0}^{t_f} \left\{ \left(\frac{\partial L^a}{\partial \mathbf{x}}\right)^T \delta\mathbf{x} + \left(\frac{\partial L^a}{\partial \dot{\mathbf{x}}}\right)^T \delta\dot{\mathbf{x}} + \left(\frac{\partial L^a}{\partial \mathbf{u}}\right)^T \delta\mathbf{u} \right\} dt + L^a|_{t_f} \delta t_f
\end{aligned} \tag{6.7}$$

The second term on the right-hand side is integrated by parts ($\int u dv = uv| - \int v du$) as follows:

$$\int_{t_0}^{t_f} \left(\frac{\partial L^a}{\partial \dot{\mathbf{x}}}\right)^T \delta\dot{\mathbf{x}} dt = \left(\frac{\partial L^a}{\partial \dot{\mathbf{x}}}\right)^T \delta\mathbf{x}|_{t_0}^{t_f} - \int_{t_0}^{t_f} \frac{d}{dt} \left(\frac{\partial L^a}{\partial \dot{\mathbf{x}}}\right)^T \delta\mathbf{x} dt$$

Then, (6.7) can be written as follows:

$$\delta J \approx \int_{t_0}^{t_f} \left\{ \left(\frac{\partial L^a}{\partial \mathbf{x}}\right)^T \delta \mathbf{x} - \frac{d}{dt}\left(\frac{\partial L^a}{\partial \dot{\mathbf{x}}}\right)^T \delta \mathbf{x} + \left(\frac{\partial L^a}{\partial \mathbf{u}}\right)^T \delta \mathbf{u} \right\} dt + \left(\frac{\partial L^a}{\partial \dot{\mathbf{x}}}\right)^T \delta \mathbf{x}\Big|_{t_0}^{t_f} + L^a|_{t_f} \delta t_f \tag{6.8}$$

Because the initial state is fixed, $\delta \mathbf{x}(t_0) = 0$, therefore

$$\delta J \approx \int_{t_0}^{t_f} \left\{ \left[\left(\frac{\partial L^a}{\partial \mathbf{x}}\right) - \frac{d}{dt}\left(\frac{\partial L^a}{\partial \dot{\mathbf{x}}}\right)\right]^T \delta \mathbf{x} + \left(\frac{\partial L^a}{\partial \mathbf{u}}\right)^T \delta \mathbf{u} \right\} dt + \left(\frac{\partial L^a}{\partial \dot{\mathbf{x}}}\right)^T_{t_f} \delta \mathbf{x}(t_f) + L^a|_{t_f} \delta t_f \tag{6.9}$$

When the cost is optimum, the increment $\delta J = 0$. Hence, the necessary conditions for optimal control are as follows: $*$ refers to the optimum.

$$\left(\frac{\partial L^a}{\partial \mathbf{x}}\right)_* - \frac{d}{dt}\left(\frac{\partial L^a}{\partial \dot{\mathbf{x}}}\right)_* = 0 \tag{6.10}$$

$$\left(\frac{\partial L^a}{\partial \boldsymbol{\lambda}}\right) - \frac{d}{dt}\left(\frac{\partial L^a}{\partial \dot{\boldsymbol{\lambda}}}\right) = 0 \rightarrow \left(\frac{\partial L^a}{\partial \boldsymbol{\lambda}}\right) = 0 \tag{6.11}$$

$$\left(\frac{\partial L^a}{\partial \mathbf{u}}\right)_* = 0 \tag{6.12}$$

$$\left(\frac{\partial L^a}{\partial \dot{\mathbf{x}}}\right)^T_* \Bigg|_{t_f} \delta \mathbf{x}(t_f) + L^a|_{t_f} \delta t_f = 0 \tag{6.13}$$

The four equations above can be described as follows: (6.10) is the Lagrange–Euler optimality criterion to be satisfied by the co-states, or the Lagrange multipliers; (6.11) presents the state dynamics $\mathbf{f}(t, \mathbf{x}(t), \mathbf{u}(t)) - \dot{\mathbf{x}} = 0$ acting as the constraint to be satisfied; (6.12) is the optimal control; and (6.13) are the boundary conditions to be satisfied. If the final state is fixed, $\delta \mathbf{x}_f = 0$, and if the final time is fixed, $\delta t_f = 0$. If the final time and final state are both fixed, the boundary condition (6.13) does not exist.

Nature of the Boundary Condition

A state trajectory for a scalar control signal is illustrated in Fig. 6.1 where it shows the state variation at t_f when the optimal control is slightly disturbed as $u(t) = u^*(t) + \delta u$. Referring to Fig. 6.1, the total end state deflection is $\delta \mathbf{x}_f = \delta \mathbf{x}(t_f) + \dot{\mathbf{x}}(t_f)\delta t_f$; hence, the boundary condition (6.13) can be re-written as follows:

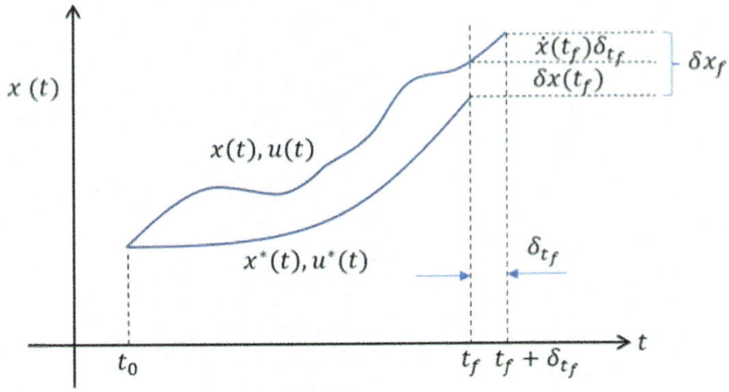

Fig. 6.1 State trajectory variation at time t_f when the optimum control signal $u^*(t)$ is slightly perturbed

$$\left(\frac{\partial L^a}{\partial \dot{\mathbf{x}}}\right)_*^T \bigg|_{t_f} \left\{\delta \mathbf{x}_f - \dot{\mathbf{x}}(t_f)\delta t_f\right\} + L^a|_{t_f}\delta t_f = 0$$

$$\left[L^a - \left(\frac{\partial L^a}{\partial \dot{\mathbf{x}}}\right)^T \dot{\mathbf{x}}\right]_{*t_f} \delta t_f + \left(\frac{\partial L^a}{\partial \dot{\mathbf{x}}}\right)_{*t_f}^T \delta \mathbf{x}_f = 0 \tag{6.14}$$

Nature of the Optimality

From (6.5) the rate of change of the cost with respect to state and control around the optimal control is as follows:

$$\frac{\partial J}{\partial \mathbf{x}^*} + \frac{\partial J}{\partial \mathbf{u}^*} = \frac{\partial}{\partial \mathbf{x}^*}\int_{t_0}^{t_f} L^a dt + \frac{\partial}{\partial \mathbf{u}^*}\int_{t_0}^{t_f} L^a dt = \int_{t_0}^{t_f}\left(\frac{\partial L^a}{\partial \mathbf{x}^*} + \frac{\partial L^a}{\partial \mathbf{u}^*}\right) dt = 0 \tag{6.15}$$

Here, differentiation and integration are interchanged, referring to the Leibniz rule of integration given in Appendix A. From the outset, it emphasizes that the integral over $[t_0, t_f]$ has to add up to zero with the possibility of having +ve and −ve values for $\frac{\partial L^a}{\partial \mathbf{x}^*}$ and $\frac{\partial L^a}{\partial \mathbf{u}^*}$ at different instances within $[t_0, t_f]$. This is a misperception. Let's take a very small control time window $[t, t + \delta t]$ for which the optimality condition (6.15) must be satisfied as follows:

$$\frac{\partial J}{\partial \mathbf{x}^*} + \frac{\partial J}{\partial \mathbf{u}^*} = \int_t^{t+\delta t}\left(\frac{\partial L^a}{\partial \mathbf{x}^*} + \frac{\partial L^a}{\partial \mathbf{u}^*}\right) dt = \frac{\partial L^a}{\partial \mathbf{x}^*}\bigg|_t \delta \mathbf{x}^* + \frac{\partial L^a}{\partial \mathbf{u}^*}\bigg|_t \delta \mathbf{u}^* = 0$$

$$\text{which requires that } \frac{\partial L^a}{\partial \mathbf{x}^*}\bigg|_t = 0 \quad \text{and} \quad \frac{\partial L^a}{\partial \mathbf{u}^*}\bigg|_t = 0 \quad \forall t \tag{6.16}$$

This verifies that the optimal control u^* optimizes the augmented Lagrangian at each $t \in [t_0, t_f]$. On the other hand, if the control signal $u(t)$ is non-optimal at any instance t, the overall control is non-optimal as well.

The second derivative of the cost with respect to state and control around the optimal control is as follows:

$$\begin{aligned}
\frac{\partial^2 J}{\partial \mathbf{x}^{*2}} + \frac{\partial^2 J}{\partial \mathbf{x}^* \partial \mathbf{u}^*} + &= \int_{t_0}^{t_f} \left(\frac{\partial^2 L^a}{\partial \mathbf{x}^{*2}} \delta^2 \mathbf{x} + \frac{\partial^2 L^a}{\partial \mathbf{x}^* \partial \mathbf{u}^*} \delta \mathbf{x} \delta \mathbf{u} + \right. \\
\frac{\partial^2 J}{\partial \mathbf{u}^* \partial \mathbf{x}^*} + \frac{\partial^2 J}{\partial \mathbf{u}^{*2}} \quad & \quad \left. \frac{\partial^2 L^a}{\partial \mathbf{u}^* \partial \mathbf{x}^*} \delta \mathbf{u} \delta \mathbf{x} + \frac{\partial^2 L^a}{\partial \mathbf{u}^{*2}} \delta^2 \mathbf{u} \right) dt \\
&= \int_{t_0}^{t_f} \begin{pmatrix} \delta \mathbf{x} & \delta \mathbf{u} \end{pmatrix} \begin{bmatrix} \frac{\partial^2 L^a}{\partial \mathbf{x}^{*2}} & \frac{\partial^2 L^a}{\partial \mathbf{x}^* \partial \mathbf{u}^*} \\ \frac{\partial^2 L^a}{\partial \mathbf{u}^* \partial \mathbf{x}^*} & \frac{\partial^2 L^a}{\partial \mathbf{u}^{*2}} \end{bmatrix} \begin{pmatrix} \delta \mathbf{x} \\ \delta \mathbf{u} \end{pmatrix} dt \\
&= \int_{t_0}^{t_f} \begin{pmatrix} \delta \mathbf{x} & \delta \mathbf{u} \end{pmatrix} \nabla_{\mathbf{x}^*, \mathbf{u}^*} L^a \begin{pmatrix} \delta \mathbf{x} \\ \delta \mathbf{u} \end{pmatrix} dt
\end{aligned} \tag{6.17}$$

where $\nabla_{\mathbf{x}^*, \mathbf{u}^*} L^a$ is the Hessian of the augmented Lagrangian. Then, the nature of the optimal control can be verified using the Hessian as follows:

$$\begin{array}{lll}
\text{if } \nabla_{\mathbf{x}^*, \mathbf{u}^*} L^a > 0 \text{ (+ve definite)} & \text{optimality is a minima} \\
\text{if } \nabla_{\mathbf{x}^*, \mathbf{u}^*} L^a < 0 \text{ (−ve definite)} & \text{optimality is a maxima} \\
\text{if } \nabla_{\mathbf{x}^*, \mathbf{u}^*} L^a = 0 \quad \text{(indefinite)} & \text{optimality is a saddle}
\end{array} \tag{6.18}$$

A few exercises on continuous-time optimal control are presented below.

Exercise: Free Final Time Optimal Control
Design the optimal control for the following plant:

$$\begin{aligned}
\dot{\mathbf{x}}(t) &= \begin{bmatrix} 0 & 1 \\ 0 & -2 \end{bmatrix} \mathbf{x}(t) + \begin{bmatrix} 0 \\ 1 \end{bmatrix} u(t) \\
&= \begin{bmatrix} x_2(t) \\ -2x_2(t) + u(t) \end{bmatrix}
\end{aligned}$$

to reach a final state $(3, 5)^T$ from an initial state $(0, 1)^T$ while minimizing the cost

$$J = \left[x_1(t_f) - 3\right]^2 + \frac{1}{2}\left[x_2(t_f) - 5\right]^2 + \frac{1}{2}\int_0^{t_f} \left[x_1^2(t) + 3x_2^2(t) + 2u^2(t)\right] dt \tag{6.19}$$

The final time t_f is free.

Solution: Lagrangian Formulation
From the given information, L, S, $\mathbf{f}$, and $\boldsymbol{\lambda}$ are as follows:

$$L = \frac{1}{2}\left[x_1^2(t) + 3x_2^2(t) + 2u^2(t)\right]$$

$$S = [x_1(t) - 3]^2 + \frac{1}{2}[x_2(t) - 5]^2$$

$$\mathbf{f} = \begin{bmatrix} x_2(t) \\ -2x_2(t) + u(t) \end{bmatrix}$$

$$\boldsymbol{\lambda} = \begin{bmatrix} \lambda_1(t) \\ \lambda_2(t) \end{bmatrix}$$

Next, the augmented Lagrangian is determined from (6.4), in which

$$\begin{aligned} \left(\frac{\partial S}{\partial \mathbf{x}}\right)^T \dot{\mathbf{x}} &= \begin{pmatrix} \frac{\partial S}{\partial x_1} & \frac{\partial S}{\partial x_2} \end{pmatrix} \begin{pmatrix} \dot{x}_1 \\ x_2 \end{pmatrix} \\ &= \frac{\partial S}{\partial x_1}\dot{x}_1 + \frac{\partial S}{\partial x_2}\dot{x}_2 \\ &= 2[x_1(t) - 3]\dot{x}_1(t) + [x_2(t) - 5]\dot{x}_2(t) \\ \left(\frac{\partial S}{\partial t}\right) &= 0 \qquad \text{because } S \text{ is not a function of } t \\ \boldsymbol{\lambda}^T(\mathbf{f} - \dot{\mathbf{x}}) &= \begin{bmatrix} \lambda_1(t) & \lambda_2(t) \end{bmatrix} \begin{bmatrix} x_2(t) - \dot{x}_1(t) \\ -2x_2(t) + u(t) - \dot{x}_2(t) \end{bmatrix} \\ &= \lambda_1(t)[x_2(t) - \dot{x}_1(t)] + \lambda_2(t)[-2x_2(t) + u(t) - \dot{x}_2(t)] \end{aligned}$$

The augmented Lagrangian is determined by referring to (6.4) as follows:

$$\begin{aligned} L^a = &\frac{1}{2}\left[x_1^2(t) + 3x_2^2(t) + 2u^2(t)\right] + 2(x_1(t) - 3)\dot{x}_1(t) + (x_2(t) - 5)\dot{x}_2(t) + \\ &\lambda_1(t)[x_2(t) - \dot{x}_1(t)] + \lambda_2(t)[-2x_2(t) + u(t) - \dot{x}_2(t)] \end{aligned} \tag{6.20}$$

Applying (6.10), the necessary conditions for optimality are obtained as follows:

$$\begin{aligned} x_1(t) + 2\dot{x}_1(t) - \frac{d}{dt}[2(x_1(t) - 3) - \lambda_1(t)] &= 0 \\ x_1(t) + \dot{\lambda}_1(t) &= 0 \\ \therefore\ \dot{\lambda}_1^*(t) &= -x_1^*(t) \end{aligned} \tag{6.21}$$

$$\begin{aligned} 3x_2(t) + \dot{x}_2(t) + \lambda_1(t) - 2\lambda_2(t) - \frac{d}{dt}(x_2(t) - 5 - \lambda_2(t)) &= 0 \\ 3x_2(t) + \lambda_1(t) - 2\lambda_2(t) + \dot{\lambda}_2(t)) &= 0 \\ \therefore\ \dot{\lambda}_2^*(t) &= -3x_2^*(t) - \lambda_1^*(t) + 2\lambda_2^*(t) \end{aligned} \tag{6.22}$$

The state equations form (6.11) are as follows:

$$\begin{aligned} x_2(t) - \dot{x}_1(t) &= 0 \\ \dot{x}_1(t) &= x_2(t) \end{aligned} \tag{6.23}$$

$$\begin{aligned} -2x_2(t) + u(t) - \dot{x}_2(t) &= 0 \\ \dot{x}_2(t) &= -2x_2(t) + u(t) \end{aligned} \tag{6.24}$$

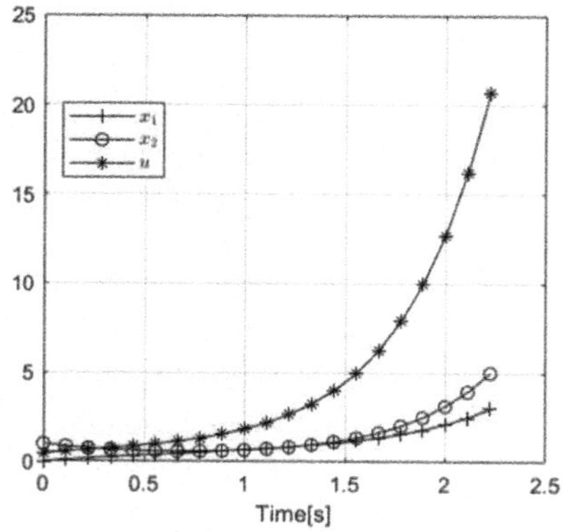

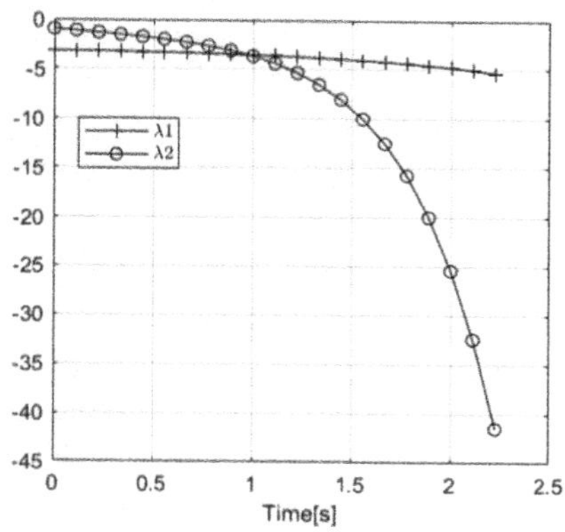

Fig. 6.2 Free final time optimal control drives the plant from the initial state $x(0) = (0, 1)^T$ to the final fixed state $x(t_f) = (3, 5)^T$. The final time is not fixed, and the optimal control takes $t_f = 2.43$ s to reach the final state with an unbounded control signal

These two differential equations are readily available from the given state-space model. Applying the necessary condition for optimal control (6.12)

$$\begin{aligned} 2u^*(t) + \lambda_2(t) &= 0 \\ \therefore u^*(t) &= -0.5\lambda_2^*(t) \end{aligned} \tag{6.25}$$

The general boundary condition (6.14) changes as follows due to the fact that $\delta x_f = 0$ because the final state is fixed.

$$\left[L^a - \left(\frac{\partial L^a}{\partial \dot{\mathbf{x}}} \right)^T \dot{\mathbf{x}} \right]_{t_f}^* = 0$$

$$\begin{aligned} \frac{1}{2}\left[x_1^2(t) + 3x_2^2(t) + 2u^2(t)\right] + 2[x_1(t) - 3]\dot{x}_1(t) + [x_2(t) - 5]\dot{x}_2(t) + \\ \lambda_1(t)[x_2(t) - \dot{x}_1(t)] + \lambda_2(t)[-2x_2(t) + u(t) - \dot{x}_2(t)] \\ -2[x_1(t) - 3 - \lambda_1(t)]\dot{x}_1 - [x_2(t) - 5 - \lambda_2(t)]\dot{x}_2 = 0 \end{aligned}$$

$$\frac{1}{2}\left[x_1^2(t) + 3x_2^2(t) + 2u^2(t)\right] + \lambda_1(t)x_2(t) + \lambda_2(t)[-2x_2(t) + u(t)] = 0 \tag{6.26}$$

The `Matlab` code in Appendix E solves the four first-order differential equations (6.21), (6.22), (6.23), and (6.24) and applies the boundary conditions. The optimum state and co-state trajectories and the optimal control signal are shown in Fig. 6.2.

Nature of Optimality

In the cost function (6.19), the terminal cost is positive for any state except $x_f \neq (3, 5)^T$. The optimal control state trajectory in Fig. 6.2 shows that the plant achieves $x_f = (3, 5)^T$ leaving no terminal cost. The running cost is a quadratic-convex function that is always positive for any $x_1(t) \neq 0, x_2(t) \neq 0$ and $u(t) \neq 0$. Therefore, the cost function is convex; hence, it has only one minima. The nature of the optimality can be assessed by checking the sign of the second variation of the augmented Lagrangian (6.20).

Nature of Control

According to the state Eq. (6.24), an excessive control signal can accelerate $x_2(t)$, which then accelerates $x_1(t)$. In order to bring the initial state $(0, 1)^T$ to the final state $(3, 5)^T$, the control signal has to move x_2 by four units while limiting the motion of x_1 to three units. This is quite an interesting maneuver, in which the control signal is kept low for about 3/4 of the time, letting both states grow slowly. And, towards the end, the exponential control action moves x_2 rapidly without giving time for x_1 to move much.

6.2 Pontryagin–Hamiltonian Formulation

Optimal control can be designed using the Hamiltonian method as well. Let's define the Hamiltonian as follows:

$$H(t, \mathbf{x}, \mathbf{u}, \boldsymbol{\lambda}) = L(t, \mathbf{x}, \mathbf{u}) + \boldsymbol{\lambda}^T \mathbf{f} \tag{6.27}$$

Then, $L(t, \mathbf{x}, \mathbf{u}) = H - \boldsymbol{\lambda}^T \mathbf{f}$ is substituted to the augmented Lagrangian in (6.4) as follows:

$$L^a = \frac{\partial}{\partial \mathbf{x}} S(t_f, \mathbf{x}_{t_f})\dot{\mathbf{x}} + \frac{\partial}{\partial t} S(t_f, \mathbf{x}_{t_f}) + H - \boldsymbol{\lambda}^T \dot{\mathbf{x}} \tag{6.28}$$

Then, the necessary condition for optimality (6.10) is written in terms of the Hamiltonian as follows:

$$\left(\frac{\partial H}{\partial \mathbf{x}}\right)_* - \frac{d}{dt}\left(\frac{\partial S}{\partial \mathbf{x}} - \boldsymbol{\lambda}\right)_* = 0$$

But, $\frac{\partial S}{\partial \mathbf{x}}$ does not depend on t hence $\frac{d}{dt}\left(\frac{\partial S}{\partial \mathbf{x}}\right) = 0$. Therefore,

$$\left(\frac{\partial H}{\partial \mathbf{x}}\right)_* + \dot{\boldsymbol{\lambda}} = 0 \tag{6.29}$$

The co-state condition (6.11) is written as follows:

$$\left(\frac{\partial H}{\partial \boldsymbol{\lambda}}\right)_* - \dot{\mathbf{x}} = 0 \tag{6.30}$$

The condition (6.12) is written in terms of the Hamiltonian as follows:

$$\left(\frac{\partial H}{\partial \mathbf{u}}\right)_* = 0 \tag{6.31}$$

And the boundary conditions (6.14) are written as follows:

$$\left[\frac{\partial S}{\partial \mathbf{x}}^T \dot{\mathbf{x}} + \frac{\partial S}{\partial t} + H - \boldsymbol{\lambda}^T \dot{\mathbf{x}} - \left(\frac{\partial S}{\partial \mathbf{x}} + \frac{\partial H}{\partial \dot{\mathbf{x}}} - \boldsymbol{\lambda}\right)^T \dot{\mathbf{x}}\right]_{t_f} \delta_{t_f} + \left(\frac{\partial S}{\partial \mathbf{x}} + \frac{\partial H}{\partial \dot{\mathbf{x}}} - \boldsymbol{\lambda}\right)^T_{t_f} \delta x_f = 0 \tag{6.32}$$

in which the first and fifth, as well as the fourth and seventh terms, get canceled. The sixth and ninth terms $\frac{\partial H}{\partial \dot{\mathbf{x}}} = 0$ because H does not depend on $\dot{x}$. Then the boundary condition in the Pontryagin–Hamiltonian formulation reduces to the following:

$$\left[\frac{\partial S}{\partial t} + H\right]_{t_f} \delta_{t_f} + \left(\frac{\partial S}{\partial \mathbf{x}} - \boldsymbol{\lambda}\right)^T_{t_f} \delta x_f = 0 \tag{6.33}$$

The four major equations of the Pontryagin–Hamiltonian formulation, named co-states (or multipliers), state (or system dynamics), optimal control, and boundary condition, are listed below.

$$\dot{\boldsymbol{\lambda}} = -\left(\frac{\partial H}{\partial \mathbf{x}}\right)_* \tag{6.34}$$

$$\dot{\mathbf{x}} = \left(\frac{\partial H}{\partial \boldsymbol{\lambda}}\right)_* \tag{6.35}$$

$$\left(\frac{\partial H}{\partial \mathbf{u}}\right)_* = 0 \tag{6.36}$$

$$\left[\frac{\partial S}{\partial t} + H\right]_{t_f} \delta_{t_f} + \left(\frac{\partial S}{\partial \mathbf{x}} - \boldsymbol{\lambda}\right)^T_{t_f} \delta x_f = 0 \tag{6.37}$$

The Pontryagin–Hamiltonian formulation is more compact compared to the Lagrangian formulation.

Nature of the Optima

The relationship in (6.28) confirms that the Lagrangian and Hamiltonian have the same partial derivatives with respect to **x** and **u**. Hence, the second derivative of the cost (6.17) will alter as follows when the augmented Lagrangian (6.28) is replaced by the Hamiltonian.

$$\begin{aligned} &\frac{\partial^2 J}{\partial \mathbf{x}^{*2}} + \frac{\partial^2 J}{\partial \mathbf{x}^* \partial \mathbf{u}^*} + = \int_{t_0}^{t_f} (\delta \mathbf{x}\ \delta \mathbf{u}) \begin{bmatrix} \frac{\partial^2 H}{\partial \mathbf{x}^{*2}} & \frac{\partial^2 H}{\partial \mathbf{x}^* \partial \mathbf{u}^*} \\ \frac{\partial^2 H}{\partial \mathbf{u}^* \partial \mathbf{x}^*} & \frac{\partial^2 H}{\partial \mathbf{u}^{*2}} \end{bmatrix} \begin{pmatrix} \delta \mathbf{x} \\ \delta \mathbf{u} \end{pmatrix} dt \\ &\frac{\partial^2 J}{\partial \mathbf{u}^* \partial \mathbf{x}^*} + \frac{\partial^2 J}{\partial \mathbf{u}^{*2}} \\ &\qquad = \int_{t_0}^{t_f} (\delta \mathbf{x}\ \delta \mathbf{u}) \nabla_{\mathbf{x}^*,\mathbf{u}^*} H \begin{pmatrix} \delta \mathbf{x} \\ \delta \mathbf{u} \end{pmatrix} dt \end{aligned} \tag{6.38}$$

Hence, the nature of optimality in the Pontryagin–Hamiltonian formulation is expressed as follows, very similar to that of the Lagrangian formulation.

$$\begin{array}{lll} \text{if } \nabla_{\mathbf{x}^*,\mathbf{u}^*} H > 0 & (+\text{ve definite}) & \text{optimality is a minima} \\ \text{if } \nabla_{\mathbf{x}^*,\mathbf{u}^*} H < 0 & (-\text{ve definite}) & \text{optimality is a maxima} \\ \text{if } \nabla_{\mathbf{x}^*,\mathbf{u}^*} H = 0 & (\text{indefinite}) & \text{optimality is a saddle} \end{array} \tag{6.39}$$

6.2.1 *The Evolution of Hamiltonian Through the Optimal Trajectory*

Let's explore how the Hamiltonian evolves through the optimal trajectory. From (6.4), the augmented Lagrangian in short-hand form is as follows:

$$L^a = \frac{\partial S}{\partial \mathbf{x}}^T \dot{\mathbf{x}} + \frac{\partial S}{\partial t} + L + \boldsymbol{\lambda}^T (\mathbf{f} - \dot{\mathbf{x}}) \tag{6.40}$$

And then,

$$\frac{\partial L^a}{\partial \dot{\mathbf{x}}} = \frac{\partial S}{\partial \mathbf{x}} - \boldsymbol{\lambda} \tag{6.41}$$

The boundary condition (6.14) is written as follows with the corresponding substitutions from (6.40) and (6.41).

$$\begin{aligned} &\left[\frac{\partial S}{\partial \mathbf{x}}^T \dot{\mathbf{x}} + \frac{\partial S}{\partial t} + L + \boldsymbol{\lambda}^T (\mathbf{f} - \dot{\mathbf{x}}) - \left(\frac{\partial S}{\partial \mathbf{x}} - \boldsymbol{\lambda}\right)^T \dot{\mathbf{x}}\right]_{t_f} \delta t_f + = 0 \\ &\qquad\qquad \left[\frac{\partial S}{\partial \mathbf{x}} - \boldsymbol{\lambda}\right]_{t_f}^T \delta \mathbf{x}_f \end{aligned} \tag{6.42}$$

The first and sixth terms, as well as the fifth and seventh terms, get canceled. Therefore,

$$\left[\frac{\partial S}{\partial t} + L + \boldsymbol{\lambda}^T \mathbf{f}\right]_{t_f} \delta t_f + \left[\frac{\partial S}{\partial \mathbf{x}} - \boldsymbol{\lambda}\right]_{t_f}^T \delta \mathbf{x}_f = 0 \tag{6.43}$$

Introducing Hamiltonian (6.27) in place of Lagrangian

$$\left[\frac{\partial S}{\partial t} + H\right]_{t_f} \delta t_f + \left[\frac{\partial S}{\partial \mathbf{x}} - \boldsymbol{\lambda}\right]_{t_f}^T \delta \mathbf{x}_f = 0 \tag{6.44}$$

If the final time t_f and the final state $\mathbf{x}_f$ are not related, then the following two conditions must be satisfied for the optimal control.

$$\left[\frac{\partial S}{\partial t} + H\right]_{t_f}^* = 0 \tag{6.45}$$

$$\left[\frac{\partial S}{\partial \mathbf{x}} - \boldsymbol{\lambda}\right]_{t_f}^* = 0 \tag{6.46}$$

If the terminal cost function $S \neq f(t)$, which is very often the case, the condition (6.45) turns to the following condition under optimal control:

$$H^*|_{t_f} = 0 \tag{6.47}$$

This is a crucial finding: for free final time optimal control, the Hamiltonian at the final time is zero. Moreover, the rate at which Hamiltonian evolution occurs is as follows:

$$\frac{dH^*(t, \mathbf{x}, \mathbf{u}, \boldsymbol{\lambda})}{dt} = \frac{\partial H^*}{\partial \mathbf{x}^*}\dot{\mathbf{x}} + \frac{\partial H^*}{\partial \mathbf{u}^*}\dot{\mathbf{u}} + \frac{\partial H^*}{\partial \boldsymbol{\lambda}^*}\dot{\boldsymbol{\lambda}} + \frac{\partial H}{\partial t} \tag{6.48}$$

in which the term $\frac{\partial H^*}{\partial t} = 0$ because $H \neq f(t)$. The Hamiltonian does not change due to t alone if $\mathbf{x}(t)$, $\boldsymbol{\lambda}(t)$, and $\mathbf{u}(t)$ do not change. Also, the term $\frac{\partial H^*}{\partial \mathbf{u}^*} = 0$ as shown in (6.36) because the optimal control optimizes the Hamiltonian. Therefore, the rate of change of the Hamiltonian reduces to the following:

$$\frac{dH^*(t, \mathbf{x}, \mathbf{u}, \boldsymbol{\lambda})}{dt} = \frac{\partial H^*}{\partial \mathbf{x}^*}\dot{\mathbf{x}} + \frac{\partial H^*}{\partial \boldsymbol{\lambda}^*}\dot{\boldsymbol{\lambda}}$$

But, from (6.34) and (6.35)

$$\frac{dH^*(t, \mathbf{x}, \mathbf{u}, \boldsymbol{\lambda})}{dt} = -\dot{\boldsymbol{\lambda}}^T \dot{\mathbf{x}} + \dot{\mathbf{x}}^T \dot{\boldsymbol{\lambda}} = 0$$

Therefore,

$$\frac{dH^*}{dt} = 0 \ \forall t \ \text{ if } t_f \text{ is free} \tag{6.49}$$

From (6.47) the Hamiltonian under optimal control is zero at the final time, and from (6.49) the rate of change of the Hamiltonian under optimal control is zero for all t. Therefore, it could be concluded that

$$H^*(t) = 0 \ \forall t \text{ if } t_f \text{ is free} \tag{6.50}$$

A few exercises on continuous-tie optimal control using the Hamiltonian are presented below.

Exercise: Free Final Time Optimal Control

Use Hamiltonian formulation and design the fixed final state and free final time optimal control for the plant presented in Sect. 6.1.

Solution:

With known L and $\mathbf{f}$ the Hamiltonian (6.27) is as follows.

$$\begin{aligned} H &= \frac{1}{2}\left[x_1^2(t) + 3x_2^2(t) + 2u^2(t)\right] + \left[\lambda_1(t)\ \lambda_2(t)\right]\begin{bmatrix} x_2(t) \\ -2x_2(t) + u(t) \end{bmatrix} \\ &= \frac{1}{2}\left[x_1^2(t) + 3x_2^2(t) + 2u^2(t)\right] + \lambda_1(t)x_2(t) + \lambda_2(t)[-2x_2(t) + u(t)] \end{aligned} \tag{6.51}$$

From (6.35)

$$\dot{\mathbf{x}} = \left(\frac{\partial H}{\partial \boldsymbol{\lambda}}\right) = \begin{bmatrix} x_2(t) \\ -2x_2(t) + u(t) \end{bmatrix} \tag{6.52}$$

which are the same state Eqs. (6.23) and (6.24) determined using the Lagrangian formulation. From (6.34),

$$\dot{\boldsymbol{\lambda}}(t) = -\left(\frac{\partial H}{\partial \mathbf{x}}\right) = \begin{bmatrix} -x_1(t) \\ -3x_2(t) - \lambda_1(t) + 2\lambda_2(t) \end{bmatrix} \tag{6.53}$$

which are the same co-state Eqs. (6.21) and (6.22) determined using the Lagrange formulation. From (6.36) the optimal control is

$$\left(\frac{\partial H}{\partial \mathbf{u}}\right) = 2u(t) + \lambda_2(t) = 0$$

$$\therefore u^*(t) = -0.5\lambda_2^*(t) \tag{6.54}$$

which is the same optimal control determined using the Lagrange formulation. Both formulations deliver the same result; however, the Hamiltonian formulation is more compact and efficient than the Lagrangian formulation.

Exercise: Free Final State Optimal Control

Design the optimal control for the plant presented in Sect. 6.1 to achieve $x_1(t_f) = 5$ in $t_f = 2$s from an initial state $\mathbf{x}(0) = (0, 1)$. Let the final state $x_2(t_f)$ be free. Use a Hamiltonian formulation.

Solution:

In this case, the functions L, S and the system dynamics $\mathbf{f}$ remain unchanged, hence the Hamiltonian (6.51) remains unchanged as well. The differential equations (6.52) for states, (6.53) for co-states, and (6.54) for optimal control also remain unchanged. The only thing that is different here is the boundary condition. When the final time is fixed, $\delta t_f = 0$, hence the general boundary condition (6.37) changes as follows:

$$\left(\frac{\partial S}{\partial \mathbf{x}} - \boldsymbol{\lambda}^T\right)_{t_f} \delta x_f = 0$$

$$\left(\frac{\partial S}{\partial x_1(t)} - \lambda_1(t) \; \frac{\partial S}{\partial x_2(t)} - \lambda_2(t)\right)^T_{t_f} \begin{pmatrix} \delta x_{1f} \\ \delta x_{2f} \end{pmatrix} = 0 \tag{6.55}$$

But, because $x_1(t_f)$ is fixed, $\delta x_{1f} = 0$, hence the boundary condition reduces as follows:

$$\left[\frac{\partial S}{\partial x_2(t)} - \lambda_2(t)\right]_{tf} \delta(x)_{2f} = 0$$

But $\delta x_{2f} \neq 0$ because x_{2f} is free. Therefore, the condition for optimality is as follows:

$$\begin{aligned} \left[\frac{\partial S}{\partial x_2(t)} - \lambda_2(t)\right]_{tf} &= 0 \\ [x_2(t) - 5 - \lambda_2(t)]_{tf} &= 0 \\ \lambda_2(t_f) &= x_2(t_f) - 5 \end{aligned} \tag{6.56}$$

The `Matlab` code in Appendix F which is a modified version of the code in Appendix E by replacing the boundary condition with (6.56) solves the optimal control problem and delivers the following result.

As shown in the left result, the desired final state $x_1(t_f) = 5$ is achieved within the set final time $t_f = 2$s. The state variable, which is left free, reaches $x_2(t_f) = 2.11$ (Fig. 6.3).

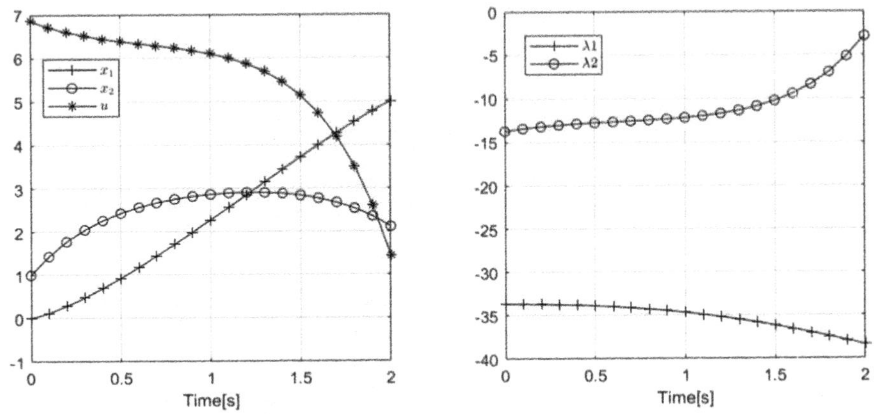

Fig. 6.3 Optimal control reaches fixed $x_1(t_f) = 5$ in fixed final time $t_f = 2$s while the free state reaches the final state $x_2(t_f) = 2.11$ under unbounded control

Exercise: Fixed Final State and Fixed Final Time Optimal Control

Use Hamiltonian formulation and design the optimal control for the plant presented in Sect. 6.1 for the final state $\mathbf{x}(t_f) = (6, 3)^T$, the final time $t_f = 3$, and the initial state $\mathbf{x}(0) = (1, 1)^T$.

Solution:
With the same functions L, S, $\mathbf{f}$, and H, the same differential equations (6.52) for states, (6.53) for co-states, and (6.54) for optimal control are valid in this case also. Because of fixed final time and fixed final state, $\delta t_f = \delta \mathbf{x}_f = 0$. Hence, the boundary

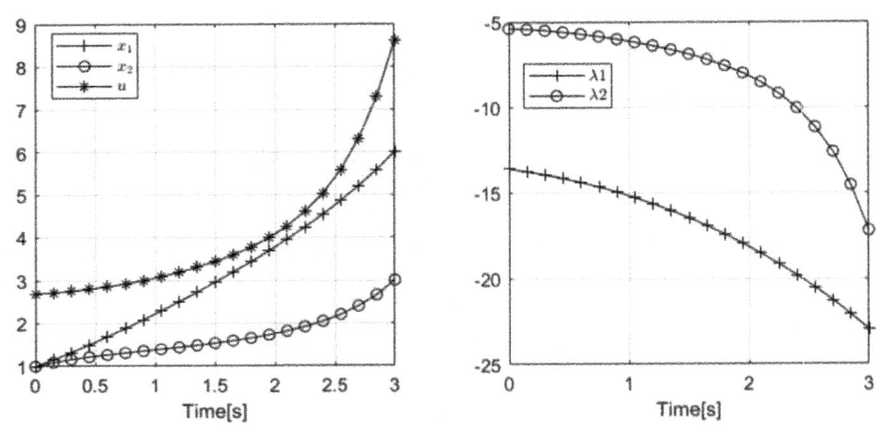

Fig. 6.4 Optimal control drives the plant from the initial state $x(0) = (1, 1)^T$ to the final state $x(0) = (3, 6)^T$ in fixed time $t_f = 3$s under unbounded control

condition (6.37) is not applicable. The initial and final states are the only applicable boundary conditions. The `Matlab` code given in Appendix G solves the differential equations with the given boundary conditions and plots the results shown in Fig. 6.4.

The optimal controls presented up to now are a process of offline programming, i.e., the control signal profile is pre-determined and applied assuming that the states undergo the expected changes. The next chapter presents how optimal control can be implemented in the form of a state feedback action, which led to the development of the famous linear quadratic regulator (LQR).

Chapter 7
Linear Quadratic Regulator

This chapter presents how optimal control is used in the form of feedback so that optimal control can be implemented on real systems. We consider linear plants and quadratic cost functions and develop the optimal regulator, which is known as the linear quadratic regulator (LQR). The nature of the optimality under LQR is analyzed, and the finite-time and infinite-time regulation processes are presented. Finally, the performance of LQR and PID control systems is compared.

7.1 Finite-Time LQR

Let's consider the following time-varying system:

$$\dot{\mathbf{x}}(t) = A(t)\mathbf{x}(t) + B\mathbf{u}(t) \tag{7.1}$$

which is to be regulated to $\mathbf{x} = 0$ in a finite time while minimizing the cost defined as follows in quadratic form

$$J = \frac{1}{2}\left\{\mathbf{x}(t_f)^T P(t_f)\mathbf{x}(t_f)\right\} + \frac{1}{2}\int_{t_0}^{t_f} \left\{\mathbf{x}(t)^T Q(t)\mathbf{x}(t) + \mathbf{u}(t)^T R(t)\mathbf{u}(t) + \boldsymbol{\lambda}^T(t)[A(t)\mathbf{x}(t) + B\mathbf{u}(t) - \dot{\mathbf{x}}(t)]\right\} dt \tag{7.2}$$

where $P(t_f)$, $Q(t)$, and $R(t)$ are symmetric positive definite matrices that convert terminal state $\mathbf{x}(t_f)$, instantaneous state $\mathbf{x}(t)$, and instantaneous control effort $\mathbf{u}(t)$ into cost. The cost defined in quadratic form confirms that it has a single optimum point, which is a minima. For a revision on quadratic functions, refer to Sect. 2.3.

S. R. Munasinghe, *Optimization and Optimal Control in a Nutshell*, Engineering Optimization: Methods and Applications,
https://doi.org/10.1007/978-981-97-8167-6_7

LQR Design

The Hamiltonian $H = L + \boldsymbol{\lambda}^T \mathbf{f}$ and the terminal cost S of the problem are as follows:

$$H = \frac{1}{2}\left\{\mathbf{x}(t)^T Q(t)\mathbf{x}(t) + \mathbf{u}(t)^T R(t)\mathbf{u}(t)\right\} + \boldsymbol{\lambda}(t)^T \left\{A(t)\mathbf{x}(t) + B\mathbf{u}(t)\right\} \quad (7.3)$$

$$S = \frac{1}{2}\mathbf{x}(t_f)^T P(t_f)\mathbf{x}(t_f) \quad (7.4)$$

The necessary condition for optimal control (6.36) is used as follows, together with the vector calculus identity (A.9) in Appendix A and the symmetric property of R.

$$\left(\frac{\partial H}{\partial \mathbf{u}}\right)_* = \frac{1}{2}(R\mathbf{u}^* + R^T\mathbf{u}^*) + B^T\boldsymbol{\lambda}^* = 0$$

Because R is symmetric

$$R\mathbf{u}^* + B^T\boldsymbol{\lambda}^* = 0$$

$$u^* = -R^{-1}B^T\boldsymbol{\lambda}^* \quad (7.5)$$

The co-state equations from (6.34) are

$$\dot{\boldsymbol{\lambda}}^* = -\left(\frac{\partial H}{\partial \mathbf{x}}\right)_* = -\frac{1}{2}\left(Q\mathbf{x} + Q^T\mathbf{x}\right) - \boldsymbol{\lambda}^{*T}A = 0$$

Because Q is symmetric

$$\dot{\boldsymbol{\lambda}}^* = -Q\mathbf{x} - A^T\boldsymbol{\lambda}^* \quad (7.6)$$

The general boundary conditions (6.37) change as follows with $\delta t_f = 0$ because the final time is fixed.

$$\left(\frac{\partial S}{\partial \mathbf{x}} - \boldsymbol{\lambda}^{*T}\right)_{t_f} = 0$$

$$P(t_f)\mathbf{x}^*(t_f) - \boldsymbol{\lambda}^*(t_f) = 0$$

$$\boldsymbol{\lambda}^*(t_f) = P(t_f)\mathbf{x}^*(t_f) \quad (7.7)$$

Optimal Control in the Form of State Feedback

The boundary condition (7.7) shows that the terminal state cost weight $P(t_f)$ relates the final optimal state $\mathbf{x}^*(t_f)$ to the final optimal co-state $\boldsymbol{\lambda}^*(t_f)$. Let's assume that the same relationship exists for any time t as follows:

$$\boldsymbol{\lambda}^*(t) = P(t)\mathbf{x}^*(t) \quad (7.8)$$

When this relationship is used in (7.5), the optimal control law takes the following shape:

$$\mathbf{u}^*(t) = -R^{-1}B^T P(t)\mathbf{x}^*(t) \tag{7.9}$$

which is a type of state feedback control. When this optimal control law is applied to (7.1), the optimal state dynamics is obtained as follows:

$$\dot{\mathbf{x}}^*(t) = (A - BR^{-1}B^T P(t))\mathbf{x}^*(t) \tag{7.10}$$

Determination of $P(t)$

Differentiating (7.8)

$$\dot{\boldsymbol{\lambda}}^* = \dot{P}\mathbf{x}^* + P\dot{\mathbf{x}}^*$$

Once substituted for $\dot{\boldsymbol{\lambda}}^*$ from (7.6), and for $\dot{\mathbf{x}}^*$ from (7.1)

$$-Q\mathbf{x}^* - A^T\boldsymbol{\lambda}^* = \dot{P}\mathbf{x}^* + P(A\mathbf{x}^* + B\mathbf{u}^*)$$

Once substituted from (7.8) for $\boldsymbol{\lambda}^*$ and from (7.5) for $\mathbf{u}^*$

$$-Q\mathbf{x}^* - A^T P\mathbf{x}^* = \dot{P}\mathbf{x}^* + PA\mathbf{x}^* - PBR^{-1}B^T\boldsymbol{\lambda}^*$$

Let's substitute for $\boldsymbol{\lambda}^*$ from (7.8)

$$-Q\mathbf{x}^* - A^T P\mathbf{x}^* = \dot{P}\mathbf{x}^* + PA\mathbf{x}^* - PBR^{-1}B^T P\mathbf{x}^*$$

Once the common $\mathbf{x}$ term is removed, the following matrix differential equation is obtained.

$$\dot{P} = -PA - A^T P - Q + PBR^{-1}B^T P \tag{7.11}$$

This Riccati equation (matrix differential equation) can be solved with known A, B, Q, R, and the boundary condition $P(t_f)$. The matrix $P(t)$ is an $n \times n$ square matrix, and it can also be verified that $P(t)$ is symmetric.

Nature of the Optimal Cost

From (7.6),

$$\begin{aligned}
\dot{\boldsymbol{\lambda}} &= -Q\mathbf{x} - A^T\boldsymbol{\lambda}^* \\
A^T\boldsymbol{\lambda}^* &= -Q\mathbf{x} - \dot{\boldsymbol{\lambda}} \\
\boldsymbol{\lambda}^{*T} A &= -\mathbf{x}^T Q - \dot{\boldsymbol{\lambda}}^T \text{ ; given } Q^T = Q
\end{aligned} \tag{7.12}$$

From (7.9) and (7.8)

$$\begin{aligned}
\mathbf{u}^*(t) &= -R^{-1}B^T\boldsymbol{\lambda}^* \\
B^T\boldsymbol{\lambda}^* &= -R\mathbf{u}^* \\
\boldsymbol{\lambda}^{*T}B &= -\mathbf{u}^{*T}R\text{ ; given } R^T = R
\end{aligned} \tag{7.13}$$

The two terms $\boldsymbol{\lambda}^T A$ and $\boldsymbol{\lambda}^T B$ in the cost (7.2) are replaced by the two equivalent expressions derived in (7.12) and (7.13) for the optimal control; hence, the optimal cost would be as follows:

$$\begin{aligned}
J^* = \frac{1}{2}\mathbf{x}^{*T}P\mathbf{x}^*\Big|_{t_f} &+ \frac{1}{2}\int_{t_0}^{t_f}\{\mathbf{x}^{*T}Q\mathbf{x}^* + \mathbf{u}^{*T}R\mathbf{u}^* + \\
&(-\mathbf{x}^{*T}Q - \dot{\boldsymbol{\lambda}}^{*T})\mathbf{x}^* - \mathbf{u}^{*T}R\mathbf{u}^* - \boldsymbol{\lambda}^{*T}\dot{\mathbf{x}}^*\Big\}\,dt \\
=& \frac{1}{2}\mathbf{x}^{*T}P\mathbf{x}^*\Big|_{t_f} - \frac{1}{2}\int_{t_0}^{t_f}(\dot{\boldsymbol{\lambda}}^{*T}\mathbf{x}^* + \boldsymbol{\lambda}^{*T}\dot{\mathbf{x}}^*)dt \\
=& \frac{1}{2}\mathbf{x}^{*T}P\mathbf{x}^*\Big|_{t_f} - \frac{1}{2}\int_{t_0}^{t_f}\frac{d}{dt}(\boldsymbol{\lambda}^{*T}\mathbf{x}^*)dt \\
=& \frac{1}{2}\mathbf{x}^{*T}P\mathbf{x}^*\Big|_{t_f} - \frac{1}{2}(\boldsymbol{\lambda}^{*T}\mathbf{x}^*)\Big|_{t_0}^{t_f} dt
\end{aligned} \tag{7.14}$$

But, from (7.8) $\boldsymbol{\lambda}^{*T} = \mathbf{x}^{*T}P$, therefore,

$$\begin{aligned}
J^* &= \frac{1}{2}\mathbf{x}^{*T}P\mathbf{x}^*\Big|_{t_f} - \frac{1}{2}(\mathbf{x}^{*T}P\mathbf{x}^*)\Big|_{t_0}^{t_f} dt \\
&= \frac{1}{2}\mathbf{x}^{*T}P\mathbf{x}^*\Big|_{t_f} - \frac{1}{2}\mathbf{x}^{*T}P\mathbf{x}^*\Big|_{t_f} + \frac{1}{2}\mathbf{x}^{*T}P\mathbf{x}^*\Big|_{t_0} \\
&= \frac{1}{2}\mathbf{x}^{*T}P\mathbf{x}^*\Big|_{t_0}
\end{aligned} \tag{7.15}$$

This is the minimum cost in finite-time optimal control, with the cost defined in quadratic form as given in (7.2). This indicates a very significant fact: the minimum cost of the entire regulation process within $[t_0, t_f]$ solely depends on the initial state $\mathbf{x}(t_0)$. Hence, it could be deduced that the minimum cost-to-go from any $t \in [t_0, t_f]$ depends solely on $\mathbf{x}(t)$ as follows:

$$J^*(t) = \frac{1}{2}\mathbf{x}^{*T}P\mathbf{x}^*\Big|_{t} \tag{7.16}$$

Hence, it could be confirmed that the optimal control minimizes the cost-to-go at each $t \in [t_0, t_f]$, which is the fundamental principle of Richard Bellman's "dynamic programming" (Bellman 1952; Szmuk et al. 2018).

LQR Design Process

The steps involved in the LQR design process are listed as follows:

1. Solve the matrix differential Riccati equation using A, B, C, Q, and R and the final state cost weight $P(t_f)$.
2. Calculate $P(t)$ for $t \in [t_0, t_f]$.
3. For each t determine the optimal state $\mathbf{x}^*$ by solving the differential equation $\dot{\mathbf{x}} = (A - BR^{-1}B^T P)\mathbf{x}$.
4. Calculate the optimal control using $\mathbf{u}^* = -R^{-1}B^T P\mathbf{x}$.

Nature of the LOR Optimality

The Hessian on the Hamiltonian (6.38) decides the nature of the optima. The Hessian, which is the matrix of second-order partial derivatives of the Hamiltonian (7.3) with respect to $\mathbf{x}$ and $\mathbf{u}$, is determined using the vector calculus identity (A.10) in Appendix A.

$$\begin{aligned}
\frac{\partial^2 H}{\partial \mathbf{x}^{*2}} &= \frac{1}{2}(Q + Q^T) = Q \text{ Because } Q \text{ is symmetric} \\
\frac{\partial^2 H}{\partial \mathbf{x}^* \partial \mathbf{u}^*} &= 0 \\
\frac{\partial^2 H}{\partial \mathbf{u}^* \partial \mathbf{x}^*} &= 0 \\
\frac{\partial^2 H}{\partial \mathbf{u}^{*2}} &= \frac{1}{2}(R + R^T) = R \text{ Because } R\text{is symmetric}
\end{aligned}$$

Then the Hessian of the Hamiltonian is as follows:

$$\nabla_{\mathbf{x}^*,\mathbf{u}^*} H = \begin{bmatrix} \frac{\partial^2 H}{\partial \mathbf{x}^{*2}} & \frac{\partial^2 H}{\partial \mathbf{x}^* \partial \mathbf{u}^*} \\ \frac{\partial^2 H}{\partial \mathbf{u}^* \partial \mathbf{x}^*} & \frac{\partial^2 H}{\partial \mathbf{u}^{*2}} \end{bmatrix} = \begin{bmatrix} Q & 0 \\ 0 & R \end{bmatrix} \tag{7.17}$$

The purpose of the two matrices Q and R is to calculate the instantaneous cost of the state $\mathbf{x}(t)$ and control $\mathbf{u}(t)$; hence, they are assumed to be symmetric and positive definite. Then, the Hessian matrix is also positive, defined definite $\nabla_{\mathbf{x}^*,\mathbf{u}^*} H > 0$. This assures the minimum cost under LQR control.

Note: To check the positive definiteness of a symmetric matrix, show that all its eigenvalues are strictly positive.

7.1.1 Finite-Time LQR Design

Exercise:

Design a finite-time LQR for the following plant

$$\dot{\mathbf{x}}(t) = \begin{bmatrix} 0 & 1 \\ -1 & 2 \end{bmatrix} \begin{pmatrix} x_1(t) \\ x_2(t) \end{pmatrix} + \begin{bmatrix} 0 \\ 1 \end{bmatrix} u(t) \tag{7.18}$$

to minimize the cost

$$J = \frac{1}{2} \left[x_1(t_f)^2 + 2x_1(t_f)x_2(t_f) + 3x_2(t_f)^2 \right] + \\ \frac{1}{2} \int_{t_0}^{t_f} \left[(x_1^2(t) + 4x_1(t)x_2(t) + 3x_2^2(t) + 0.5u^2(t) \right] dt \tag{7.19}$$

considering initial state $\mathbf{x}(0) = (2, -1)^T$ and final time $t_f = 8$s.

Solution

The plant starts with $\mathbf{x}(0) = (2, -1)^T$, and the optimal control is expected to regulate it, i.e., to bring the state to $\mathbf{x}(t_f) = (0, 0)^T$ within the fixed final time $t_f = 8$s while minimizing the cost. The given system dynamics reveal the system matrix A and drive vector B as follows:

$$A = \begin{bmatrix} 0 & 1 \\ -1 & 2 \end{bmatrix} \quad B = \begin{bmatrix} 0 \\ 1 \end{bmatrix}$$

The cost function is arranged as follows:

$$J = \frac{1}{2} \mathbf{x}(t_f)^T \begin{bmatrix} 1 & 1 \\ 1 & 3 \end{bmatrix} \mathbf{x}(t_f) + \frac{1}{2} \int_{t_0}^{t_f} \mathbf{x}(t)^T \begin{bmatrix} 1 & 2 \\ 2 & 3 \end{bmatrix} \mathbf{x}(t) + \\ u(t)^T 0.5 u(t) dt \tag{7.20}$$

Then, the matrices $P(t_f)$, Q, and R are revealed as follows:

$$P(t_f) = \begin{bmatrix} 1 & 1 \\ 1 & 3 \end{bmatrix} \quad Q = \begin{bmatrix} 1 & 2 \\ 2 & 3 \end{bmatrix} \quad R = 0.5$$

With known A, B, Q, and R, the matrix differential Riccati equation (7.11) is determined to be the following:

$$\begin{bmatrix} \dot{p}_{11} & \dot{p}_{12} \\ \dot{p}_{21} & \dot{p}_{22} \end{bmatrix} = - \begin{bmatrix} p_{11} & p_{12} \\ p_{21} & p_{22} \end{bmatrix} \begin{bmatrix} 0 & 1 \\ -1 & 2 \end{bmatrix} - \begin{bmatrix} 0 & -1 \\ 1 & 2 \end{bmatrix} \begin{bmatrix} p_{11} & p_{12} \\ p_{21} & p_{22} \end{bmatrix} - \begin{bmatrix} 1 & 2 \\ 2 & 3 \end{bmatrix} + \\ \begin{bmatrix} p_{11} & p_{12} \\ p_{21} & p_{22} \end{bmatrix} \begin{bmatrix} 0 \\ 1 \end{bmatrix} 2 \begin{bmatrix} 0 & 1 \end{bmatrix} \begin{bmatrix} p_{11} & p_{12} \\ p_{21} & p_{22} \end{bmatrix}$$

The Riccati equation can be expanded to the following four differential equations: For each differential equation, the final value is known from $P(t_f)$.

$$\begin{aligned}
\dot{p}_{11} &= p_{12} + p_{21} - 1 + 2p_{12}p_{21} \; ; \; p_{11}(t_f) = 1 \\
\dot{p}_{12} &= -p_{11} - 2p_{12} + p_{22} - 2 + 2p_{12}p_{22} \; ; \; p_{12}(t_f) = 1 \\
\dot{p}_{21} &= p_{22} - p_{11} - 2p_{21} - 2 + 2p_{21}p_{22} \; ; \; p_{21}(t_f) = 1 \\
\dot{p}_{22} &= -p_{21} - 4p_{22} - p_{12} - 3 + 2p_{22}^2 \; ; \; p_{22}(t_f) = 3
\end{aligned}$$

It is noticeable that $\dot{p}_{12} = \dot{p}_{21}$ because $P(t)$ is symmetric. Hence, there are only three differential equations existing.

The `Matlab` code in Appendix H solves these three differential equations and calculates the optimal control (7.9), optimal state trajectories (7.10), and the cost-to-go elements in $P(t)$. As evident, $p_{12}(t) = p_{21}(t)$ because $P(t)$ is symmetric. It is intriguing to witness that the $P(t)$ matrix elements stay unchanged for the most part and start to change only towards the end. Due to the fact that $P(t)$ depends on time, true state feedback is not possible in the finite-time LQR.

7.2 Infinite-Time LQR

In infinite-time control, it is assumed that $\mathbf{x}(t_f) \to 0$ as $t_f \to \infty$ making the terminal cost zero. Then, the cost function (7.2) will change as follows:

$$J = \frac{1}{2}\int_{t_0}^{t_f} \left[\mathbf{x}(t)^T Q\mathbf{x}(t) + \mathbf{u}(t)^T R\mathbf{u}(t)\right] dt \tag{7.21}$$

As shown in Fig. 7.1, the cost-to-go matrix elements have a zero rate of change right throughout, except towards the very end of the fixed final time. If the final time is relaxed, the control becomes infinite-time, and it would be reasonable to assume that $\dot{P}(t) = 0 \forall t \in [t_0, t_f \to \infty]$, and therefore $P(t)$ becomes a constant matrix. This assumption has a very significant outcome because when $\dot{P}(t) = 0$ the differential Riccati equation (7.11) will become an algebraic Riccati equation, as shown below.

$$PA + A^T P + Q - PBR^{-1}B^T P = 0 \tag{7.22}$$

This algebraic Riccati equation is solved to determine the constant P matrix, which is substituted in (7.9) to determine the infinite-time LQR control as follows:

$$\begin{aligned}
\mathbf{u}^*(t) &= -R^{-1}B^T P\mathbf{x}^*(t) \\
&= -K_{LQR}\mathbf{x}^*(t)
\end{aligned} \tag{7.23}$$

where K_{LQR} is the $m \times n$ constant LQR gain matrix for a plant with n-states and m-control inputs.

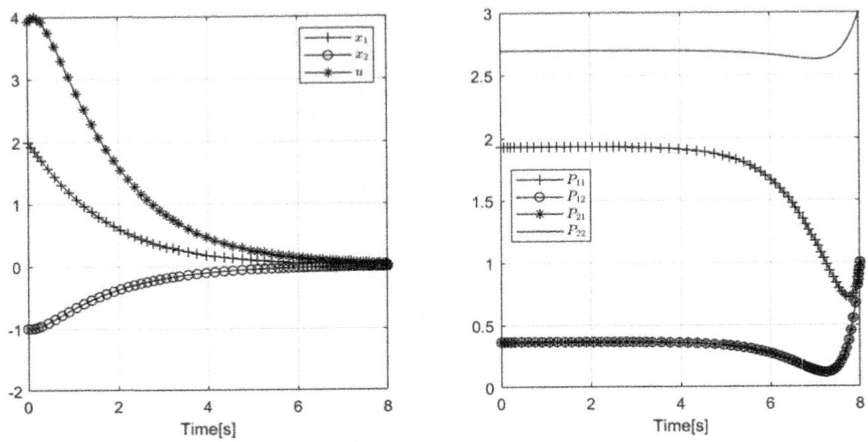

Fig. 7.1 Finite-time LQR regulates the plant from its initial state of $(-1, 2)^T$ within the final time $t_f = 8$s

7.2.1 *Infinite-Time LQR Design*

Exercise:

Design an infinite-time LQR for the same plant presented in Sect. 7.1.1.

$$\dot{\mathbf{x}}(t) = \begin{bmatrix} 0 & 1 \\ -1 & 2 \end{bmatrix} \begin{pmatrix} x_1(t) \\ x_2(t) \end{pmatrix} + \begin{bmatrix} 0 \\ 1 \end{bmatrix} u(t) \tag{7.24}$$

Using the same cost

$$J = \frac{1}{2} \int_{t_0}^{\infty} \left[(x_1^2(t) + 2x_1(t)x_2(t) + 3x_2^2(t) + 0.5u^2(t) \right] dt \tag{7.25}$$

and the same initial state, $\mathbf{x}(0) = (2, -1)^T$.

Solution:

For this plant and cost function, the following matrices have been determined in Sect. 7.1.1.

$$A = \begin{bmatrix} 0 & 1 \\ -1 & 2 \end{bmatrix} \quad B = \begin{bmatrix} 0 \\ 1 \end{bmatrix} \quad Q = \begin{bmatrix} 1 & 2 \\ 2 & 3 \end{bmatrix} \quad R = 0.5 \tag{7.26}$$

Then, the algebraic Riccati equation (7.22) with $p_{12} = p_{21} = p$ is as follows:

$$PA + A^T P + Q - PBR^{-1}B^T P = 0$$
$$\begin{bmatrix} p_{11} & p \\ p & p_{22} \end{bmatrix}\begin{bmatrix} 0 & 1 \\ -1 & 2 \end{bmatrix} + \begin{bmatrix} 0 & -1 \\ 1 & 2 \end{bmatrix}\begin{bmatrix} p_{11} & p \\ p & p_{22} \end{bmatrix} + = 0$$
$$\begin{bmatrix} 1 & 2 \\ 2 & 3 \end{bmatrix} - \begin{bmatrix} p_{11} & p \\ p & p_{22} \end{bmatrix}\begin{bmatrix} 0 \\ 1 \end{bmatrix} 2 \begin{bmatrix} 0 & 1 \end{bmatrix}\begin{bmatrix} p_{11} & p \\ p & p_{22} \end{bmatrix}$$
$$\begin{bmatrix} -p & p_{11} + 2p \\ -p_{22} & p + 2p_{22} \end{bmatrix} + \begin{bmatrix} -p & -p_{22} \\ p_{11} + 2p & p + 2p_{22} \end{bmatrix} + \begin{bmatrix} 1 & 2 \\ 2 & 3 \end{bmatrix} - \begin{bmatrix} p^2 & pp_{22} \\ pp_{22} & p_{22}^2 \end{bmatrix} = 0$$

which is resolved into the following three nonlinear equations:

$$p^2 + 2p - 1 = 0 \tag{7.27}$$
$$p_{22}^2 - 4p_{22} - 2p - 3 = 0 \tag{7.28}$$
$$p_{11} + 2p - p_{22}(1 + p) + 2 = 0 \tag{7.29}$$

Once solved, the solution space of p, p_{11} and p_{22} is as follows:

p(7.27)	p_{22}(7.28)	p_{11}(7.29)	Remarks
−1.618	1.872	−2.954	$p_{11} < 0 \rightarrow P$ is not +ve def
−1.618	0.126	0.954	
0.618	3.288	4.116	
0.618	−1.288		$p_{22} < 0 \rightarrow P$ is not +ve def

Then, the two possible candidates for P are as follows:

$$P = \begin{bmatrix} 0.954 & -1.618 \\ -1.618 & 0.126 \end{bmatrix} \text{ and } P = \begin{bmatrix} 4.116 & 0.618 \\ 0.618 & 3.288 \end{bmatrix} \tag{7.30}$$

The second principle minors of the two matrices are –2.5 and 13.151. Hence, the second matrix is the one that is +ve definite. With this P matrix, the infinite-time LQR (7.23) is written as follows:

$$u^*(t) = -2\begin{bmatrix} 0 & 1 \end{bmatrix}\begin{bmatrix} 4.116 & 0.618 \\ 0.618 & 3.288 \end{bmatrix}\mathbf{x}(t)$$
$$= -\begin{bmatrix} 0.618 & 3.288 \end{bmatrix}\mathbf{x}(t) \tag{7.31}$$

The constant gains make the real-time LQR possible through state feedback.

LQR and PID Comparison

The two simulink models shown in Fig. 7.2 use the same plant presented in Sect. 7.2.1 with known A, B, C, D. The left figure shows the infinite-time LQR with $K_{LQR} = -[0.618, 3.288]$, and the figure on the right shows the PID controller.

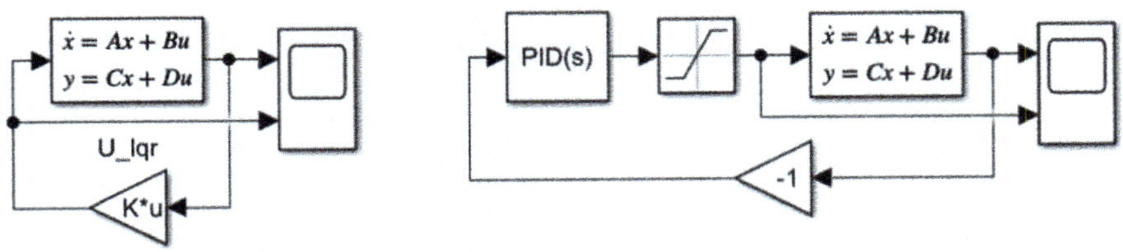

Fig. 7.2 LQR and PID control

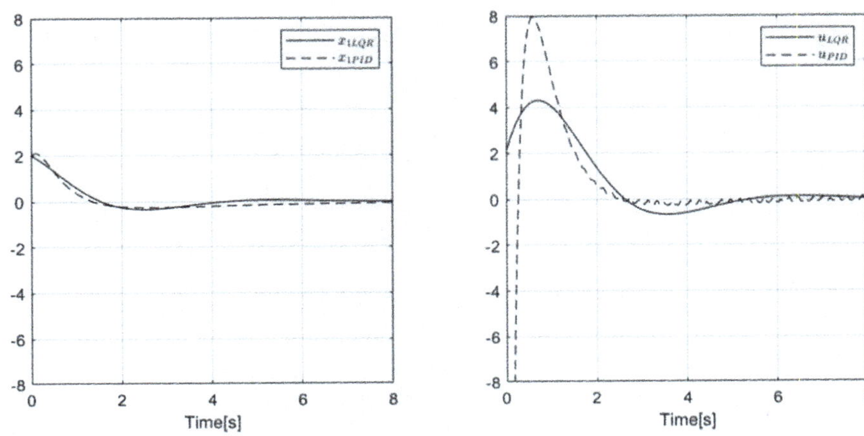

Fig. 7.3 State regulation and control signals of LQR and PID control

The PID controller gains tuned in `Matlab` are $K_P = 369$, $K_I = 820$, and $K_D = 41$. These gains generate a huge control signal, and to keep it within comparable limits of the LQR, the PID gains are adjusted to $K_P = 8$, $K_I = 8$, and $K_D = 2$. Both simulations start with the same initial conditions: $(2, -1)^T$.

Figure 7.3 shows the regulation performance of LQR and PID controls. It is obvious that the LQR brings the state $x_1(t)$ to the origin much better than the PID controller, which shows a more oscillatory response. It is also evident that the LQR maintains a small and less aggressive control signal compared to the PID control signal. The `Matlab` code in Appendix I implements this simulation together with the Simulink model shown in Fig. 7.2.

The next chapter presents Pontryagin's maximum principle, which is used when the optimal criterion is to maximize performance while minimizing cost.

References

Bellman RE (1952) The theory of dynamic programming. In: Proceedings of the national academy of science, vol 38, p 716

Szmuk M, Noyan N, Muharremoglu A (2018) A dynamic programming algorithm for mixed-integer linear-quadratic control problems. IEEE Trans Autom Control 63(12):4135–4140

Chapter 8
Pontryagin's Maximum Principle

Some control systems need to maximize a certain outcome while minimizing the cost. For example, a rocket propulsion system intends to reach the maximum altitude with the least fuel. In such systems, the cost minimization concept in optimal control is changed to outcome maximization, and the particular optimal control design process can be implemented using Pontryagin's maximum principle. This chapter presents the maximum principle based on the Hamiltonian and demonstrates its application through several exercises. The bang–bang control and control switching, which are outcomes of the maximum principle, are also presented.

8.1 Maximizing a Performance Index

When it is expected to maximize the net outcome while minimizing the cost, the performance function is defined as follows:

$$J = \Gamma(\mathbf{x}(t_f)) - \int_{t_0}^{t_f} \left\{ L(t, \mathbf{x}, \mathbf{u}) + \boldsymbol{\lambda}^T (\dot{\mathbf{x}} - \mathbf{f}(\mathbf{x}, \mathbf{u}) \right\} dt \tag{8.1}$$

where $\Gamma(\mathbf{x}(t_f))$ is the net performance at the final state $\mathbf{x}(t_f)$ and the augmented Lagrangian $L^a = L + \boldsymbol{\lambda}^T (\dot{\mathbf{x}} - \mathbf{f})$ is the instantaneous cost. The objective is to maximize the net performance J. Let's define the Hamiltonian differently so that the performance and Hamiltonian have the same signs.

$$H = \boldsymbol{\lambda}^T \mathbf{f} - L \tag{8.2}$$

S. R. Munasinghe, *Optimization and Optimal Control in a Nutshell*, Engineering Optimization: Methods and Applications,
https://doi.org/10.1007/978-981-97-8167-6_8

When the Lagrangian L in (8.1) is replaced by the Hamiltonian in (8.2), the net performance is as follows:

$$\begin{aligned} J &= \Gamma(\mathbf{x}(t_f)) - \int_{t_0}^{t_f} \left\{\boldsymbol{\lambda}^T \mathbf{f} - H + \boldsymbol{\lambda}^T(\dot{\mathbf{x}} - \mathbf{f})\right\} dt \\ &= \Gamma(\mathbf{x}(t_f)) - \int_{t_0}^{t_f} \left\{\boldsymbol{\lambda}^T \dot{\mathbf{x}} - H\right\} dt \end{aligned} \tag{8.3}$$

The optimal performance J^* is achieved with the optimal control $\mathbf{u}^*$ through the optimal state trajectory $\mathbf{x}^*$. If the control is perturbed slightly from its optimal the performance will degrade. Hence, the following inequality holds:

$$\begin{aligned} J(\mathbf{u}^*) \geq& J(\mathbf{u}^* + \delta\mathbf{u}) \\ \Gamma(\mathbf{x}^*(t_f)) - \int_{t_0}^{t_f} \left\{\boldsymbol{\lambda}^T \dot{\mathbf{x}}^* - H_{\mathbf{u}^*\mathbf{x}^*}\right\} dt \geq& \Gamma(\mathbf{x}^*(t_f)) + \frac{\partial \Gamma(\mathbf{x}^*(t_f))^T}{\partial \mathbf{x}} \delta\mathbf{x} - \\ & \int_{t_0}^{t_f} \left\{\boldsymbol{\lambda}^T(\dot{\mathbf{x}}^* + \delta\dot{\mathbf{x}}) - H_{\mathbf{u}^*+\delta\mathbf{u},\mathbf{x}^*+\delta\mathbf{x}}\right\} dt \\ \int_{t_0}^{t_f} H_{\mathbf{u}^*\mathbf{x}^*} dt \geq& \frac{\partial \Gamma(\mathbf{x}^*(t_f))^T}{\partial \mathbf{x}} \delta\mathbf{x} + \int_{t_0}^{t_f} \left\{H_{\mathbf{u}^*+\delta\mathbf{u},\mathbf{x}^*+\delta\mathbf{x}} - \right. \\ & \left. \boldsymbol{\lambda}^T \delta\dot{\mathbf{x}}\right\} dt \end{aligned} \tag{8.4}$$

The Hamiltonian $H_{\mathbf{u}^*+\delta\mathbf{u},\mathbf{x}^*+\delta\mathbf{x}^*}$ can be approximated with the first-order Taylor expansion as follows:

$$H_{\mathbf{u}^*+\delta\mathbf{u},\mathbf{x}^*+\delta\mathbf{x}^*} = H_{\mathbf{u}^*+\delta\mathbf{u},\mathbf{x}^*} + \frac{\partial H}{\partial \mathbf{x}}^T \delta\mathbf{x} \tag{8.5}$$

and the last term in (8.4) can be evaluated as follows:

$$\begin{aligned} \int_{t_0}^{t_f} \boldsymbol{\lambda}^T \delta\dot{\mathbf{x}} &= \boldsymbol{\lambda}^T \delta\mathbf{x}\Big|_{t_0}^{t_f} - \int_{t_0}^{t_f} \dot{\boldsymbol{\lambda}}^T \delta\mathbf{x} dt \\ &= \boldsymbol{\lambda}(t_f)^T \delta\mathbf{x} - \int_{t_0}^{t_f} \dot{\boldsymbol{\lambda}}^T \delta\mathbf{x} dt \end{aligned} \tag{8.6}$$

Let's substitute from (8.5) and (8.6) into (8.4) to obtain the following:

$$\begin{aligned} \int_{t_0}^{t_f} H_{\mathbf{u}^*\mathbf{x}^*} dt \geq& \frac{\partial \Gamma(\mathbf{x}^*(t_f))^T}{\partial \mathbf{x}} \delta\mathbf{x} + \int_{t_0}^{t_f} \left\{H_{\mathbf{u}^*+\delta\mathbf{u},\mathbf{x}^*} + \frac{\partial H}{\partial \mathbf{x}}^T \delta\mathbf{x} + \dot{\boldsymbol{\lambda}}^T \delta\mathbf{x}\right\} dt \\ & - \boldsymbol{\lambda}(t_f)^T \delta\mathbf{x} \end{aligned}$$

When the co-state equation $\dot{\boldsymbol{\lambda}}(t) = -\frac{\partial H}{\partial \mathbf{x}}$ in the Hamiltonian formulation is applied, the inequality reduces as follows:

$$\int_{t_0}^{t_f} H_{\mathbf{u}^*\mathbf{x}^*} dt \geq \frac{\partial \Gamma(\mathbf{x}^*(t_f))^T}{\partial \mathbf{x}} \delta \mathbf{x} + \int_{t_0}^{t_f} H_{\mathbf{u}^*\delta \mathbf{u},\mathbf{x}^*} dt - \boldsymbol{\lambda}(t_f)^T \delta \mathbf{x} \tag{8.7}$$

If the following boundary conditions are enforced for the Lagrange multipliers

$$\boldsymbol{\lambda}(t_f) = \frac{\partial \Gamma(\mathbf{x}^*(t_f))}{\partial \mathbf{x}} \tag{8.8}$$

The condition (8.7) for maximum performance is as follows:

$$\int_{t_0}^{t_f} H_{\mathbf{u}^*\mathbf{x}^*} dt \geq \int_{t_0}^{t_f} H_{\mathbf{u}^*+\delta \mathbf{u},\mathbf{x}^*} dt \tag{8.9}$$

For any arbitrary t and an infinitesimal time window $[t - \delta t, t + \delta t]$, the following statement must be true for the maximum net outcome.

$$\int_{t-\delta t}^{t+\delta t} H_{\mathbf{u}^*\mathbf{x}^*}(t) dt \geq \int_{t-\delta t}^{t+\delta t} H_{\mathbf{u}^*+\delta \mathbf{u},\mathbf{x}^*}(t) dt \tag{8.10}$$

Therefore, the sufficient condition for the maximum net outcome is as follows:

$$H_{\mathbf{u}^*\mathbf{x}^*}(t) \geq H_{\mathbf{u}^*+\delta \mathbf{u},\mathbf{x}^*}(t) \qquad \forall t \tag{8.11}$$

This inequality is known as Pontryagin's maximum principle.

8.1.1 Maximum Distance with Minimum Energy

Exercise:
Figure 8.1 shows a mobile robot being pushed by a control signal u.

Determine the optimal control that maximizes the distance z in time t_f while minimizing the energy consumption as defined by the following performance function.

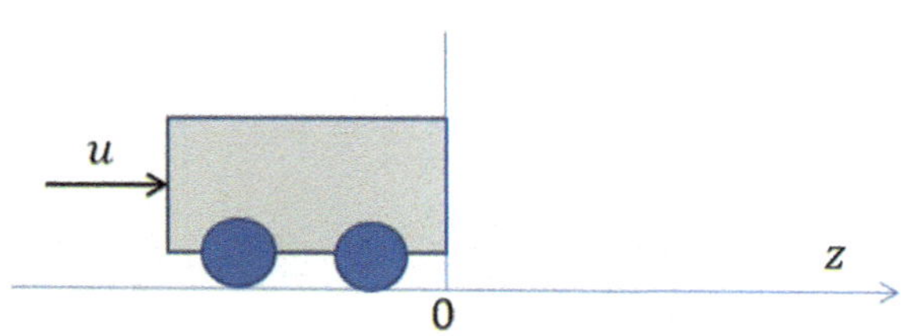

Fig. 8.1 Mobile robot being moved by a control signal

$$J(u) = z(T) - \int_0^{t_f} \frac{1}{2}u^2(t)dt \tag{8.12}$$

Solution:

Let's define the state variables as $x_1 = z$(distance) and $\dot{x}_1 = x_2 = \dot{z}$(velocity). Then, the state-space model of the robot is as follows:

$$\mathbf{f} = \begin{bmatrix} f_1 \\ f_2 \end{bmatrix} = \begin{pmatrix} \dot{x}_1 \\ \dot{x}_2 \end{pmatrix} = \begin{bmatrix} 0 & 1 \\ 0 & 0 \end{bmatrix} \begin{pmatrix} x_1 \\ x_2 \end{pmatrix} + \begin{bmatrix} 0 \\ 1 \end{bmatrix} u = \begin{pmatrix} x_2 \\ u \end{pmatrix} \tag{8.13}$$

in which the control signal is considered the acceleration. $\dot{x}_2 = u$. From (8.12), the Lagrangian of the control problem is

$$L = \frac{1}{2}u^2(t) \tag{8.14}$$

From (8.2), the Hamiltonian is determined as follows:

$$\begin{aligned} H(t) &= \begin{bmatrix} \lambda_1(t) & \lambda_2(t) \end{bmatrix} \begin{bmatrix} x_2(t) \\ u(t) \end{bmatrix} - \frac{1}{2}u^2(t) \\ &= \lambda_1(t)x_2(t) + \lambda_2(t)u(t) - \frac{1}{2}u^2(t) \end{aligned} \tag{8.15}$$

As derived in (8.11), the control is optimal when the Hamiltonian is maximized with respect to the control. Therefore, the optimal control is determined as follows:

$$\begin{aligned} \frac{\partial H(t)}{\partial u^*(t)} &= \lambda_2(t) - u^*(t) = 0 \\ \text{therefore, } u^*(t) &= \lambda_2(t) \end{aligned} \tag{8.16}$$

Comparing the performance functions (8.3) and (8.12) the terminal benefit to be maximized is

$$\Gamma(\mathbf{x}(t_f)) = z(t_f) = x_1(t_f) \tag{8.17}$$

Then, from $\boldsymbol{\lambda}(t_f) = \frac{\partial \Gamma(\mathbf{x}^*(t_f))}{\partial \mathbf{x}}$ in (8.8),

$$\begin{aligned} \begin{bmatrix} \lambda_1(t_f) \\ \lambda_2(t_f) \end{bmatrix} &= \begin{bmatrix} \frac{\partial \Gamma(x_1(t_f))}{\partial x_1} \\ \frac{\partial \Gamma(x_1(t_f))}{\partial x_2} \end{bmatrix} \\ &= \begin{bmatrix} 1 \\ 0 \end{bmatrix} \end{aligned} \tag{8.18}$$

From the co-state equation $\dot{\boldsymbol{\lambda}}(t) = -\frac{\partial H}{\partial \mathbf{x}}$, the rate of change of Lagrange multipliers is determined as follows:

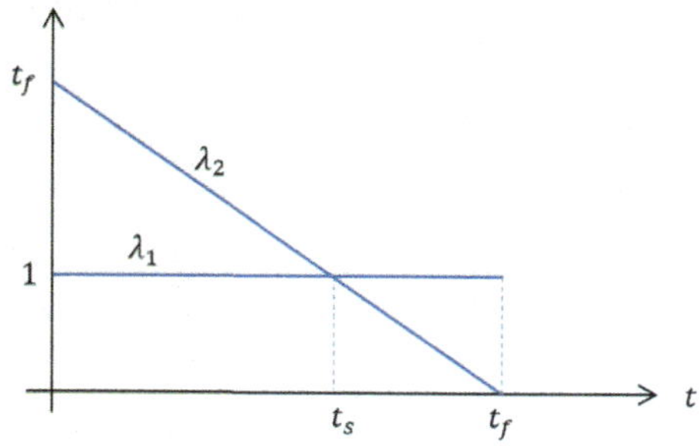

Fig. 8.2 Trajectories of the two Lagrange multipliers $\lambda_1(t)$ and $\lambda_2(t)$

$$\begin{bmatrix} \dot{\lambda}_1(t) \\ \dot{\lambda}_2(t) \end{bmatrix} = \begin{bmatrix} 0 \\ -\lambda_1(t) \end{bmatrix} \tag{8.19}$$

From (8.18) and (8.19), $\lambda_1(t) = 1 \ \ \forall t \in [0, t_f]$ because $\dot{\lambda}_1(t) = 0$ and $\lambda_2(t) = t_f - t$ as it must drop to 0 with $-\lambda_1(t) = -1$ gradient in time t_f. Therefore, the trajectories of the two Lagrange multipliers are as follows:

$$\begin{bmatrix} \lambda_1(t) \\ \lambda_2(t) \end{bmatrix} = \begin{bmatrix} 1 \\ t_f - t \end{bmatrix} \tag{8.20}$$

These two trajectories are illustrated in Fig. 8.2. With the known $\lambda_2(t)$ in (8.20), the optimal control (8.16) can be re-written as a function of time as follows:

$$u^*(t) = \lambda_2(t) = t_f - t \tag{8.21}$$

With this optimal control, which is the robot's acceleration, the velocity $x_2(t)$ and distance $x_1(t)$ are determined as follows by successive integration of u^* assuming zero initial speed and zero initial distance.

$$\begin{bmatrix} x_1(t) \\ x_2(t) \end{bmatrix} = \begin{bmatrix} t_f \frac{1}{2} t^2 - \frac{t^3}{6} \\ t_f t - \frac{1}{2} t^2 \end{bmatrix} \tag{8.22}$$

The Hamiltonian (8.15) under the optimal control (8.21) is as follows:

$$H^* = x_2(t) + \frac{1}{2}(t_f - t)^2 \tag{8.23}$$

Figure 8.3 shows the optimal control (8.21), Hamiltonian (8.15), speed, and distance (8.22) of the robot.

In this example, the Hamiltonian is a quadratic function of u and the control signal is unbounded; therefore, it is quite straightforward to determine the optimum

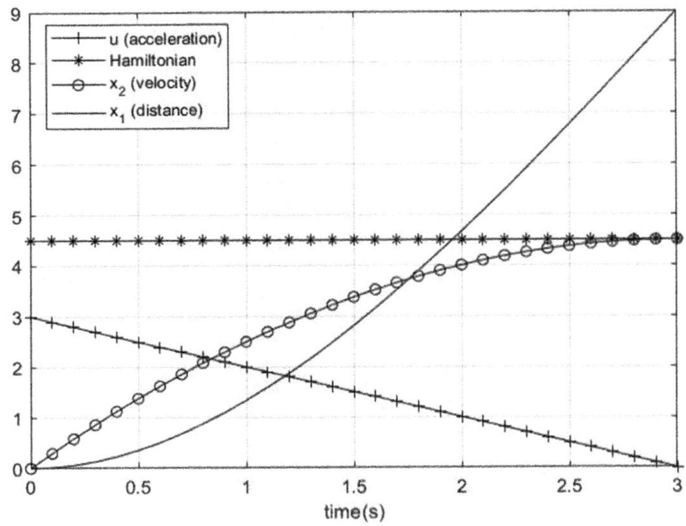

Fig. 8.3 Maximum distance and minimum energy optimal control

u by setting $\nabla_u H = 0$. However, there are systems where the Hamiltonian is non-quadratic on u. The following example shows how Pontryagin's maximum principle is used to determine an optimal control for such systems.

8.1.2 Maximum Distance with Minimum Fuel (Bang–Bang)

Let's consider the previous example (double integrator) for minimum fuel consumption as specified in the following performance function, which is non-quadratic. Consider that the control signal is bounded as $u \in [0, u_{max}]$.

$$J(u) = z(t_f) - \int_0^{t_f} u(t)dt \tag{8.24}$$

The Lagrangian in this case is

$$L = u(t) \tag{8.25}$$

State equations $\mathbf{f}$ remain unchanged. The Hamiltonian is as follows:

$$\begin{aligned} H(t, \mathbf{x}, \mathbf{u}) &= \boldsymbol{\lambda}^T \mathbf{f} - L \\ &= \begin{bmatrix} \lambda_1(t) \; \lambda_2(t) \end{bmatrix} \begin{bmatrix} x_2(t) \\ u(t) \end{bmatrix} - u(t) \\ &= \lambda_1(t)x_2(t) + (\lambda_2(t) - 1)u(t) \end{aligned} \tag{8.26}$$

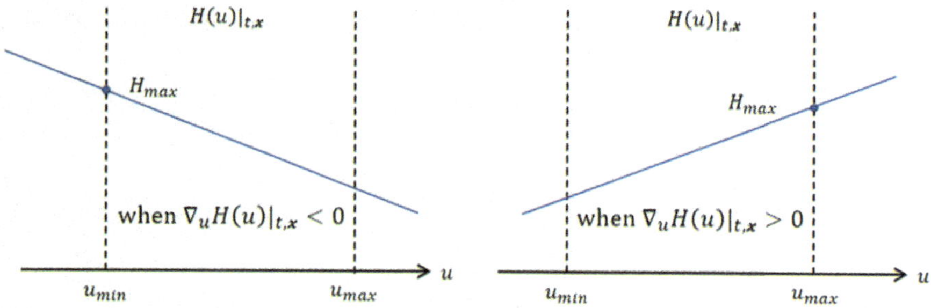

Fig. 8.4 Maximizing Hamiltonian by adjusting the control signal to its minimum or maximum limit depending on the sign of $\nabla_u H$ (Bang–Bang control)

The linear nature of the Hamiltonian does not allow the FONC $\nabla_u H(u) = 0$ to be used to determine the maximum Hamiltonian. The gradient of the Hamiltonian

$$\nabla_u H = \lambda_2(t) - 1 \tag{8.27}$$

indicates that when $\lambda_2(t) > 1 \rightarrow \nabla_u H > 0$, the necessary control to maximize H is $u = u_{max}$. On the other hand, when $\lambda_2(t) < 1 \rightarrow \nabla_u H < 0$, then the necessary control to maximize H is $u = u_{min}$. Figure 8.4 illustrates these two scenarios, and the optimal control law is formulated as follows:

$$u^*(t) = \begin{cases} u_{max} \; if \; \nabla_u H > 0 \text{ i. e. } \lambda_2(t) > 1 \\ u_{min} \; if \; \nabla_u H < 0 \text{ i. e. } \lambda_2(t) < 1 \end{cases} \tag{8.28}$$

The optimal control signal is always set to either the upper limit or the lower limit based on the value of $\lambda_2(t)$, hence the term "Bang–Bang". From the objective function (8.24), the terminal performance is $\Gamma(\mathbf{x}(t_f)) = z(t_f) = x_1(t_f)$. Then, from (8.8), the terminal values of Lagrange multipliers are determined as follows:

$$\begin{bmatrix} \lambda_1(t_f) \\ \lambda_2(t_f) \end{bmatrix} = \begin{bmatrix} \frac{\partial \Gamma(x_1(t_f))}{\partial x_1} \\ \frac{\partial \Gamma(x_1(t_f))}{\partial x_2} \end{bmatrix} = \begin{bmatrix} 1 \\ 0 \end{bmatrix} \tag{8.29}$$

From the co-state equation $\dot{\boldsymbol{\lambda}}(t) = -\frac{\partial H}{\partial \mathbf{x}}$, the rate of change of Lagrange multipliers for the given Hamiltonian (8.26) is determined as follows:

$$\begin{bmatrix} \dot{\lambda}_1(t) \\ \dot{\lambda}_2(t) \end{bmatrix} = \begin{bmatrix} 0 \\ -\lambda_1(t) \end{bmatrix} \tag{8.30}$$

From (8.29) and (8.30), the trajectories of the Lagrange multipliers are determined as follows:

$$\begin{bmatrix} \lambda_1(t) \\ \lambda_2(t) \end{bmatrix} = \begin{bmatrix} 1 \\ t_f - t \end{bmatrix} \tag{8.31}$$

Using $\lambda_2(t)$ trajectory, the optimal control criterion (8.28) can be re-written by replacing $\lambda_2(t)$ with time as follows:

$$u^* = \begin{cases} u_{max} & \forall t \in [0, t_f - 1] \\ 0 & \forall t \in [t_f - 1, t_f] \end{cases} \tag{8.32}$$

The robot uses maximum acceleration for $t_f - 1$ duration and shuts down the thrust at that moment. Then, the robot's motion is derived as follows:

$$\begin{bmatrix} x_1^*(t) \\ x_2^*(t) \end{bmatrix} = \begin{cases} \frac{1}{2} u_{max} t^2 & \forall t \in [0, t_f - 1] \\ u_{max} t & \text{- do-} \\ x_1(t_f - 1) + x_2(t_f - 1)(t - t_f + 1) & \forall t \in [t_f - 1, t_f] \\ x_2(t_f - 1) & \text{- do-} \end{cases} \tag{8.33}$$

The Hamiltonian (8.26) under optimal control (8.31) and (8.32) is as follows:

$$H^* = \begin{cases} x_2(t) + (t_f - t - 1)u_{max} & \forall t] \in [0, t_f - 1] \\ x_2(t_f - 1) & \forall t] \in [t_f - 1, t_f] \end{cases} \tag{8.34}$$

Figure 8.5 illustrates the Lagrange multiplier trajectories and how the optimal control signal switches at $t = t_f - 1$ where $\lambda_2(t) < 1$ (Fig. 8.6).

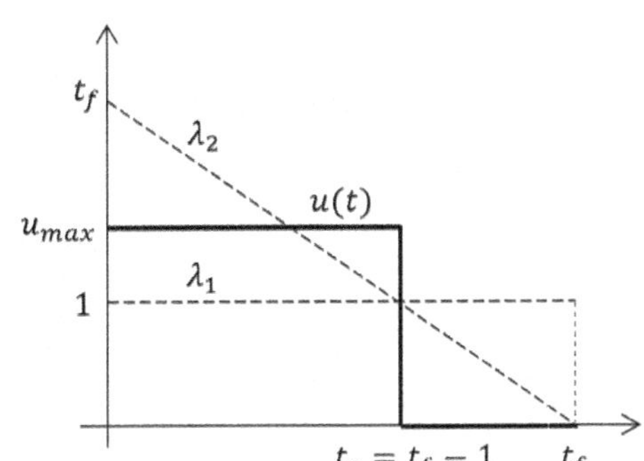

Fig. 8.5 The optimal bang–bang control

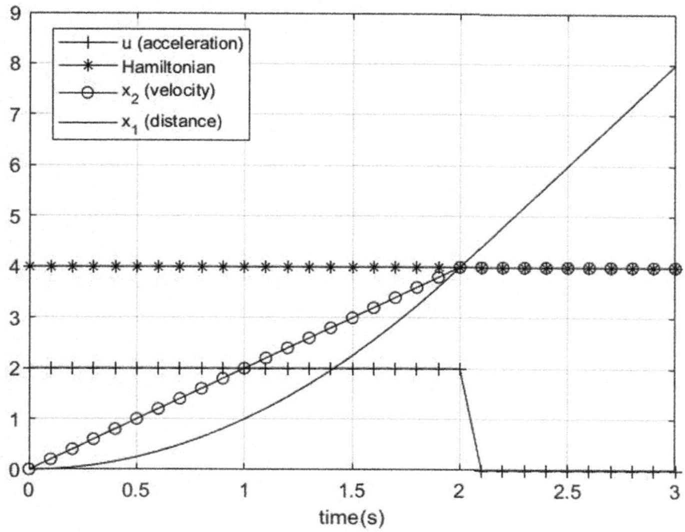

Fig. 8.6 Maximum distance and minimum energy optimal control

8.1.3 *Minimum-Time Control*

Exercise:

Design a minimum-time regulator for the following system with bounded control of $u \in [-1, +1]$.

$$\dot{\mathbf{x}} = \begin{bmatrix} 0 & 1 \\ 0 & 0 \end{bmatrix} \mathbf{x} + \begin{bmatrix} 0 \\ 1 \end{bmatrix} u \tag{8.35}$$

Demonstrate the state regulation, followed by an initial disturbance. $\mathbf{x}(0) = (2, -1)^T$.

Solution:

From the system dynamics (8.35),

$$\mathbf{f} = \begin{bmatrix} \dot{x}_1(t) \\ \dot{x}_2(t) \end{bmatrix} = \begin{bmatrix} x_2(t) \\ u(t) \end{bmatrix} \tag{8.36}$$

Minimum-time control does not have a final performance term to be maximized but only the cost, i.e., time duration, to be minimized. Therefore, the performance function for time minimization is defined as follows:

$$J = 0 - \int_{t_0}^{t_f} 1 dt$$

where $\Gamma(\mathbf{x}(t_f)) = 0$ and $L = 1$. The Hamiltonian (8.2) is as follows:

$$
\begin{aligned}
H &= \boldsymbol{\lambda}^T \mathbf{f} - L \\
&= [\lambda_1 \; \lambda_2] \begin{bmatrix} x_2 \\ u \end{bmatrix} + 1 \\
&= \lambda_1 x_2 + \lambda_2 u + 1
\end{aligned}
$$

Because the Hamiltonian is linear, as in the previous exercise, the optimal control is bang–bang, where only the two extreme controls $u \in \{u_{max}, u_{min}\}$ are used with a single switching within the control duration.

From (8.36), the state trajectory on the $x_1 - x_2$ plane is determined as follows:

$$
\begin{aligned}
\dot{x}_2 = u \rightarrow x_2 = & \quad ut + x_2(t_0); t = \tfrac{x_2 - x_2(t_0)}{u} \\
\dot{x}_1 = x_2 \rightarrow x_1 = & \quad \tfrac{1}{2}ut^2 + x_2(t_0)t + x_1(t_0) \\
= & \; \tfrac{1}{2}u(\tfrac{x_2 - x_2(t_0)}{u})^2 + x_2(t_0)(\tfrac{x_2 - x_2(0)}{u}) + x_1(t_0) \\
= & \quad \tfrac{(x_2 - x_2(t_0))^2}{2u} + x_2(t_0)(\tfrac{x_2 - x_2(t_0)}{u}) + x_1(t_0)
\end{aligned}
\tag{8.37}
$$

It turns out to be a parabolic state trajectory positioned by the initial disturbance. In fact, there are two such parabolas, as shown below, for the two extreme values of u. For the given initial disturbance $\mathbf{x}(t_0) = (2, -1)^T$, the two parabolas are as follows:

$$
x_1(t) = \begin{cases} 0.5x_2^2 + 1.5 \;\; ; u^* = +1 \\ -0.5x_2^2 + 2.5 \; ; u^* = -1 \end{cases} \tag{8.38}
$$

Figure 8.7 shows these two parabolas labeled $u = +1$ and $u = -1$. One of the points of intersection of these two state trajectories is the initial state $\mathbf{x}(t_0)$. Based on the optimal control signal $u^* = +1$ or $u^* = -1$, the state trajectory takes off along the corresponding parabola.

Regulator Development

Let's bring in the two special parabolic state trajectories that pass through (0, 0) under extreme controls. By setting $x_1(t_0) = x_2(t_0) = 0$ in (8.37), the common equation of these two parabolas is $x_1 = \frac{1}{2u}x_2^2$, and for each value of $u \in [-1, +1]$, the following two parabolas are obtained.

$$
x_1 = \begin{cases} 0.5x_2^2 \;\; \text{s–1} \; - \circ - \; \text{for} \;\; u = +1 \\ -0.5x_2^2 \; \text{s+1} \; - \diamond - \; \text{for} \;\; u = -1 \end{cases} \tag{8.39}
$$

Figure 8.7 shows these two half-parabolas labeled switching curve s–1 and switching curve s+1 using $\diamond$ and $\circ$ markers. The two switching curves meet at the origin.

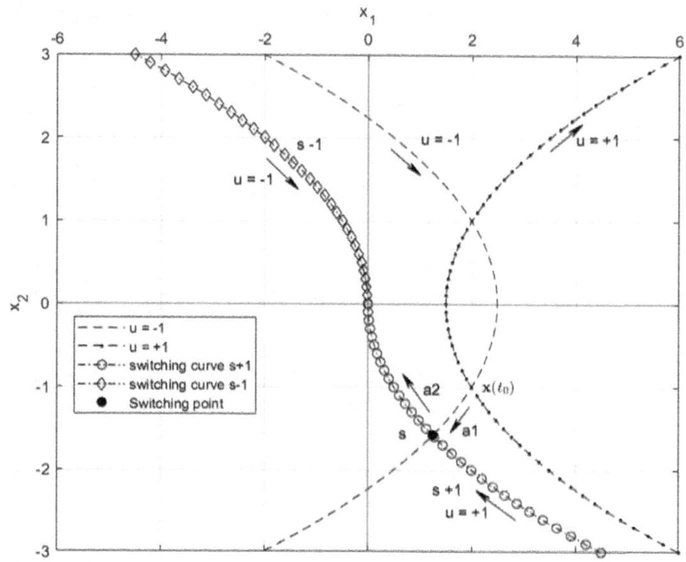

Fig. 8.7 Minimum-time control: the state takes off from $\mathbf{x}(t_0)$ with $u = -1$ and meets the switching curve at "s" where the control switches to $u = +1$ to get on the switching curve s+1 that will bring the state to the origin

Optimal Controller Design

Then, the minimum-time control procedure can be described as follows: Based on the initial state $\mathbf{x}(t_0)$, apply $u^* = -1$ initially so that the state will start moving from $\mathbf{x}(t_0)$ on the parabola as indicated by the arrow a1. When the state "s" marked • is reached, switch the control as $u = +1$ to get on the switching curve s+1 and start moving as indicated by the arrow a2 towards the origin. Because the state always travels along the parabolas constrained by the system dynamics (8.35), the controls used are the maximum allowed, and also only one switching is used, the time taken to reach the origin is the minimum possible. The control switching criterion is presented next.

Optimum Switching Criterion

Figure 8.8 shows the two sides of the state space partitioned by the combined switching curve. Our objective is to find a valid criterion across the entire state space to set the extreme control $u^* \in \{-1, +1\}$ in order to drive any initial state $\mathbf{x}(t_0)$ through a parabolic trajectory towards the closest switching curve s–1 or s+1 so that any arbitrary initial state can be regulated to $(0, 0)$ with just one switching of the control signal.

Referring to the direction of travel as shown in Fig. 8.7, when the state is to the right of the combined switching curve, $u = -1$ must be chosen to reach s+1 switching curve, whereas when the state is to the left of the combined switching curve, $u = +1$

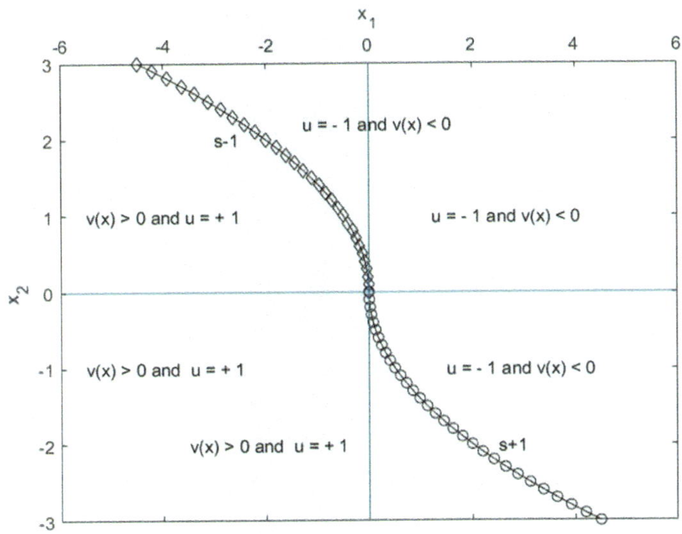

Fig. 8.8 Optimum control signal based on the state

must be chosen to reach s–1 switching curve. Hence, the control criterion just requires a method to determine whether the state is to the left or to the right of the combined switching curve. As illustrated in Fig. 8.8, the sign of the function

$$v(\mathbf{x}(t)) = -0.5x_2(t)|x_2(t)| - x_1(t) \tag{8.40}$$

agrees with the sign of the desired control signal across the entire state space. Hence, the control switching criterion can be formulated as follows:

$$u^* = \begin{cases} u_{max} = -1 \text{ if } v(\mathbf{x}(t)) < 0 \\ u_{min} = +1 \text{ if } v(\mathbf{x}(t)) > 0 \end{cases} \tag{8.41}$$

Figure 8.9 shows the Simulink block and the results of the control system.

The result shows how the extreme controls $u^* \in [-1, +1]$ drive the initial perturbation $(2, -1)$ to $(0, 0)$ with just one control switching around 0.6s. The total time taken for registration is about 2.1 s, which is the minimum time possible.

Practical Difficulty in Control Switching

Sudden switching of extreme control signals may not be acceptable to plants and systems if it causes too much stress on the actuators and physical hardware. Physical systems often find it difficult or unable to implement control switching successfully. Moreover, the chattering phenomenon, which is the high-frequency switching due to noise while the state slides along the switching curve, is also a common problem in this type of control.

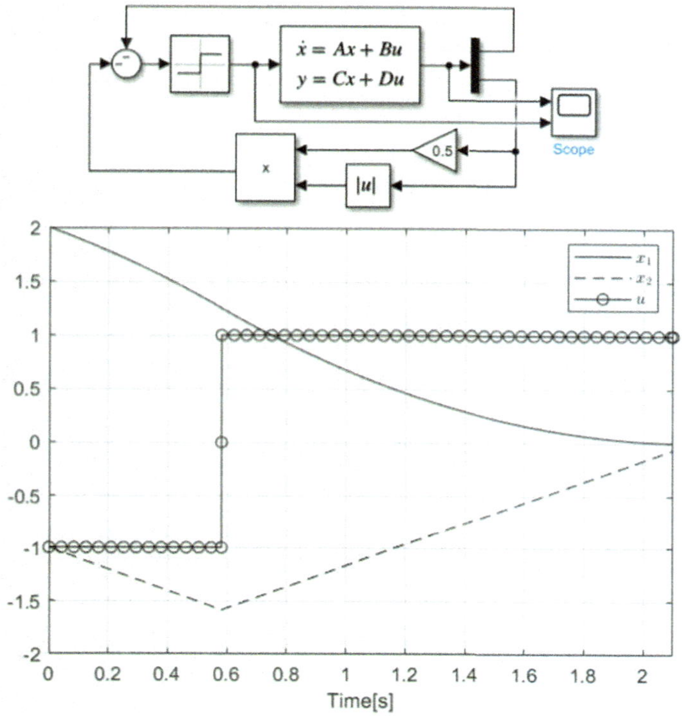

Fig. 8.9 The minimum-time bang–bang regulator and its performance

The next chapter presents how optimal control is implemented on discrete-time systems using the discrete-time state-space model and quadratic cost.

Chapter 9
Discrete-Time Optimal Control

Modern control systems are mostly implemented in discrete time. Hence, the optimal control process also needs discrete-time implementation. This chapter presents how discrete-time optimal control can be developed for discrete state-space systems with quadratic cost functions. Several worked exercises are presented to show how discrete-time optimal control is implemented.

9.1 Discrete-Time Systems

Figure 9.1 shows how a single state x transits from an initial state x_0 to a final state x_N under a series of discrete controls u_k; $k = 0, 1, 2, ...(N-1)$.

The determination of optimal control sequence involves optimization of a cost function $\mathbf{J}(k)$ along the transition $k \in [0, N]$ subject to system dynamics.

9.2 Open-Loop Unconstrained Finite-Horizon Discrete Linear Optimal Control with Quadratic Cost

Let's consider the linear, discrete system described by

$$\mathbf{x}(k+1) = \mathbf{A}\mathbf{x}(k) + \mathbf{B}\mathbf{u}(k) \tag{9.1}$$

where $\mathbf{x}(k), \mathbf{u}(k)$; $k = 0, 1, 2...N$ are $n \times 1$ state vector and $m \times 1$ control vector, $\mathbf{A}_{n \times n}$ is the $n \times n$ system matrix, and $\mathbf{B}$ is $n \times m$ drive matrix. The control objective of the regulator is to achieve $\mathbf{x}(N) = 0$ within the finite horizon of N steps, and

S. R. Munasinghe, *Optimization and Optimal Control in a Nutshell*, Engineering Optimization: Methods and Applications,
https://doi.org/10.1007/978-981-97-8167-6_9

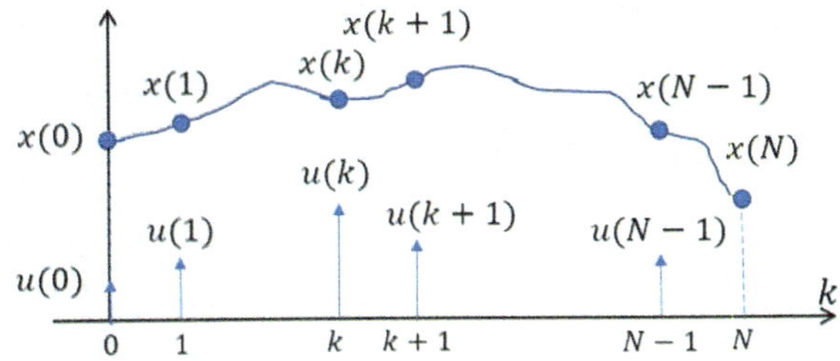

Fig. 9.1 A single state transition from x_0 to x_N under discrete control sequence u_k

therefore, any $\mathbf{x}(k) \neq 0$ is considered a cost that should be minimized. Control signals are also a cost that needs to be minimized. Hence, the instantaneous cost at the kth instance is defined as follows:

$$\mathbf{J}(k) = \frac{1}{2}\mathbf{x}(k)^T\mathbf{Q}\mathbf{x}(k) + \frac{1}{2}\boldsymbol{u}(k)^T\mathbf{R}\boldsymbol{u}(k) \tag{9.2}$$

where $\mathbf{Q}_{n\times n}$ is the diagonal matrix of error cost weights, and $\mathbf{R}_{m\times m}$ is the diagonal matrix of control cost weights. The total cost (scalar) of the control sequence $k = 0...N$ is as follows:

$$\mathbf{J} = \frac{1}{2}\sum_{k=1}^{N}\mathbf{x}(k)^T\mathbf{Q}\mathbf{x}(k) + \frac{1}{2}\sum_{k=0}^{N-1}\mathbf{u}(k)^T\mathbf{R}\mathbf{u}(k) + \frac{1}{2}\mathbf{x}(0)^T\mathbf{Q}\mathbf{x}(0) \tag{9.3}$$

The optimal control problem is then to find the control sequence $\mathbf{u}^*$ that minimizes the cost (9.3) subject to the constraint (9.1). The total cost in (9.3) is expanded as follows:

$$\begin{aligned}
\mathbf{J} = {} & \frac{1}{2}\left[\mathbf{x}(1)^T\ \mathbf{x}(2)^T\ ...\ \mathbf{x}(N)^T\right]\begin{bmatrix}\mathbf{Q} & & ... & \\ & \mathbf{Q} & ... & \\ & \vdots & & \\ & & ... & \mathbf{Q}\end{bmatrix}\begin{bmatrix}\mathbf{x}(1)\\ \mathbf{x}(2)\\ \vdots\\ \mathbf{x}(N)\end{bmatrix} + \\
& \frac{1}{2}\left[\mathbf{u}(0)^T\ \mathbf{u}(1)^T\ ...\ \mathbf{u}(N-1)^T\right]\begin{bmatrix}\mathbf{R} & & ... & \\ & \mathbf{R} & ... & \\ & \vdots & & \\ & & ... & \mathbf{R}\end{bmatrix}\begin{bmatrix}\mathbf{u}(0)\\ \mathbf{u}(1)\\ \vdots\\ \mathbf{u}(N-1)\end{bmatrix} + \\
& \frac{1}{2}\mathbf{x}(0)^T\mathbf{Q}\mathbf{x}(0) \\
= {} & \frac{1}{2}\left[\mathbf{x}^T\bar{\mathbf{Q}}\mathbf{x} + \mathbf{u}^T\bar{\mathbf{R}}\mathbf{u} + \mathbf{x}(0)^T\mathbf{Q}\mathbf{x}(0)\right]
\end{aligned} \tag{9.4}$$

where $\bar{\mathbf{Q}}$ is $nN \times nN$ matrix, $\bar{\mathbf{R}}$ is $mN \times mN$ matrix, and $\mathbf{x} = [\mathbf{x}(1)\mathbf{x}(2)..\mathbf{x}(N)]$ is a $nN \times 1$ vector, and $\mathbf{u} = [\mathbf{u}(0)\mathbf{u}(1)..\mathbf{u}(N-1)]$ is a $mN \times 1$ vector. The system dynamics (9.1) through the sequence $k = 1, 2...N$ can be written as follows:

$$\begin{aligned}
\mathbf{x}(1) &= \mathbf{A}\mathbf{x}(0) + \mathbf{B}\mathbf{u}(0) \\
\mathbf{x}(2) &= \mathbf{A}^2\mathbf{x}(0) + \begin{bmatrix}\mathbf{AB} \ \mathbf{B}\end{bmatrix}\begin{bmatrix}\mathbf{u}(0)\\ \mathbf{u}(1)\end{bmatrix} \leftarrow \mathbf{A}\mathbf{x}(1) + \mathbf{B}\mathbf{u}(1) \\
\mathbf{x}(3) &= \mathbf{A}^3\mathbf{x}(0) + \begin{bmatrix}\mathbf{A^2B} \ \mathbf{AB} \ \mathbf{B}\end{bmatrix}\begin{bmatrix}\mathbf{u}(0)\\ \mathbf{u}(1)\\ \mathbf{u}(2)\end{bmatrix} \\
&\vdots \\
\mathbf{x}(N) &= \mathbf{A}^N\mathbf{x}(0) + \begin{bmatrix}\mathbf{A}^{N-1}\mathbf{B} \ \mathbf{A}^{N-2}\mathbf{B} \ \ldots \ \mathbf{AB} \ \mathbf{B}\end{bmatrix}\begin{bmatrix}\mathbf{u}(0)\\ \mathbf{u}(1)\\ \vdots \\ \mathbf{u}(N-2)\\ \mathbf{u}(N-1)\end{bmatrix}
\end{aligned} \tag{9.5}$$

Then, all of the states are described using only the initial state $\mathbf{x}(0)$ and the control sequence up to that state as follows:

$$\begin{aligned}
\begin{bmatrix}\mathbf{x}(1)\\ \mathbf{x}(2)\\ \mathbf{x}(3)\\ \vdots\\ \mathbf{x}(N)\end{bmatrix} &= \begin{bmatrix}\mathbf{A}\\ \mathbf{A}^2\\ \mathbf{A}^3\\ \vdots\\ \mathbf{A}^N\end{bmatrix}\mathbf{x}(0) + \begin{bmatrix}\mathbf{B} & 0 & 0 & \ldots 0\\ \mathbf{AB} & \mathbf{B} & 0 & \ldots 0\\ \mathbf{A^2B} & \mathbf{AB} & \mathbf{B} & \ldots 0\\ & & & \vdots \\ \mathbf{A}^{N-1}\mathbf{B} & \mathbf{A}^{N-2}\mathbf{B} & \mathbf{A}^{N-3}\mathbf{B} & \ldots \mathbf{B}\end{bmatrix}\begin{bmatrix}\mathbf{u}(0)\\ \mathbf{u}(1)\\ \mathbf{u}(2)\\ \vdots\\ \mathbf{u}(N-1)\end{bmatrix} \\
\mathbf{x} &= \bar{\mathbf{A}}\mathbf{x}(0) + \bar{\mathbf{B}}\mathbf{u}
\end{aligned} \tag{9.6}$$

where $\bar{\mathbf{A}}$ is a $nN \times n$ matrix and $\bar{\mathbf{B}}$ is a $nN \times mN$ matrix. Using this representation, the individual states $\mathbf{x}$ in the cost function (9.4) can be replaced by the initial state $\mathbf{x}(0)$ and the control sequence $\mathbf{u}$ as follows:

$$\begin{aligned}
\mathbf{J}(\mathbf{x}(0), \mathbf{u}) &= \frac{1}{2}\left[\left(\bar{\mathbf{A}}\mathbf{x}(0) + \bar{\mathbf{B}}\mathbf{u}\right)^T \bar{\mathbf{Q}}\left(\bar{\mathbf{A}}\mathbf{x}(0) + \bar{\mathbf{B}}\mathbf{u}\right) + \mathbf{u}^T\bar{\mathbf{R}}\mathbf{u} + \mathbf{x}(0)^T\mathbf{Q}\mathbf{x}(0)\right] \\
&= \frac{1}{2}\left[\mathbf{x}(0)^T\bar{\mathbf{A}}^T\bar{\mathbf{Q}}\bar{\mathbf{A}}\mathbf{x}(0) + 2\mathbf{x}(0)^T\bar{\mathbf{A}}^T\bar{\mathbf{Q}}\bar{\mathbf{B}}\mathbf{u} + \mathbf{u}^T\bar{\mathbf{B}}^T\bar{\mathbf{Q}}\bar{\mathbf{B}}\mathbf{u} + \mathbf{u}^T\bar{\mathbf{R}}\mathbf{u} + \mathbf{x}(0)^T\mathbf{Q}\mathbf{x}(0)\right] \\
&= \frac{1}{2}\left[\mathbf{x}(0)^T\bar{\mathbf{A}}^T\bar{\mathbf{Q}}\bar{\mathbf{A}}\mathbf{x}(0) + \mathbf{x}(0)^T\mathbf{Q}\mathbf{x}(0) + 2\mathbf{x}(0)^T\bar{\mathbf{A}}^T\bar{\mathbf{Q}}\bar{\mathbf{B}}\mathbf{u} + \mathbf{u}^T\bar{\mathbf{B}}^T\bar{\mathbf{Q}}\bar{\mathbf{B}}\mathbf{u} + \mathbf{u}^T\bar{\mathbf{R}}\mathbf{u}\right] \\
&= \frac{1}{2}\mathbf{x}(0)^T\left(\bar{\mathbf{A}}^T\bar{\mathbf{Q}}\bar{\mathbf{A}} + \mathbf{Q}\right)\mathbf{x}(0) + \frac{1}{2}\mathbf{u}^T\left(\bar{\mathbf{B}}^T\bar{\mathbf{Q}}\bar{\mathbf{B}} + \bar{\mathbf{R}}\right)\mathbf{u} + \mathbf{x}(0)^T\bar{\mathbf{A}}^T\bar{\mathbf{Q}}\bar{\mathbf{B}}\mathbf{u} \\
&= \text{constant}(\mathbf{x}(0)) + \frac{1}{2}\mathbf{u}^T\bar{\mathbf{H}}\mathbf{u} + \mathbf{x}(0)^T\bar{\mathbf{F}}^T\mathbf{u}
\end{aligned} \tag{9.7}$$

where $\bar{\mathbf{H}} = \bar{\mathbf{B}}^T\bar{\mathbf{Q}}\bar{\mathbf{B}} + \bar{\mathbf{R}}$ and $\bar{\mathbf{F}}^T = \bar{\mathbf{A}}^T\bar{\mathbf{Q}}\bar{\mathbf{B}}$. The dimensions of $\bar{\mathbf{H}}$ are $[mN \times nN][nN \times nN][nN \times mN] = mN \times mN$, and the dimensions of $\bar{\mathbf{F}}^T$ are $[n \times nN][nN \times nN][nN \times mN] = n \times mN$. Then, the condition for optimal cost with respect to the control sequence is as follows:

$$\nabla_{\mathbf{u}}\mathbf{J} = \bar{\mathbf{H}}\mathbf{u} + \bar{\mathbf{F}}\mathbf{x}(0) = 0 \tag{9.8}$$

which is of the same shape as the FONC for quadratic function optimization (2.12). Then, the optimal control sequence is as follows:

$$\mathbf{u}^* = -\bar{\mathbf{H}}^{-1}\bar{\mathbf{F}}\mathbf{x}(0) \tag{9.9}$$

This is the optimal open-loop control sequence given the quadratic cost function (9.3) constrained by the system dynamics (9.1). The optimal control sequence u^* looks very similar to the optimal solution of a quadratic function derived in (2.13).

Exercise

The state-space model of a second-order plant is as follows:

$$\begin{aligned}\dot{\mathbf{x}}(t) &= \begin{bmatrix} 0 & 1 \\ -2 & -1 \end{bmatrix}\mathbf{x}(t) + \begin{bmatrix} 0 \\ 1 \end{bmatrix}u(t) \\ \mathbf{y}(t) &= \begin{bmatrix} 1 & 0 \\ 0 & 1 \end{bmatrix}\mathbf{x}(t) + \begin{bmatrix} 0 \\ 0 \end{bmatrix}u(t)\end{aligned} \tag{9.10}$$

Determine the discrete-time open-loop optimal control sequence for the following cost function:

$$\mathbf{J}(k) = x_1^2(k) + x_2^2(k) + 0.3u^2(k) \tag{9.11}$$

Simulate the plant for 2 s with a 0.2 s sampling interval and an initial state of $\mathbf{x}(0) = [0.3, -0.3]^T$.

Solution

From the continuous-time state-space model (9.10), the discrete-time state-space model for a 0.2 s sampling interval is determined as follows:

$$\begin{aligned}\mathbf{x}(k+1) &= \begin{bmatrix} 0.963 & 0.179 \\ -0.358 & 0.784 \end{bmatrix}\mathbf{x}(k) + \begin{bmatrix} 0.019 \\ 0.179 \end{bmatrix}u(k) \ ; \quad \mathbf{x}(0) = [0.3, -0.3]^T \\ \mathbf{y}(k) &= \begin{bmatrix} 1 & 0 \\ 0 & 1 \end{bmatrix}\mathbf{x}(k) + \begin{bmatrix} 0 \\ 0 \end{bmatrix}u(t)\end{aligned}$$

from which the matrices **A** and **B** of the discrete-time model (9.1) are obtained. Then, the matrices $\bar{\mathbf{A}}$, $\bar{\mathbf{B}}$ are calculated from (9.6). From the given cost function (9.11), **Q** and **R** can be read as follows:

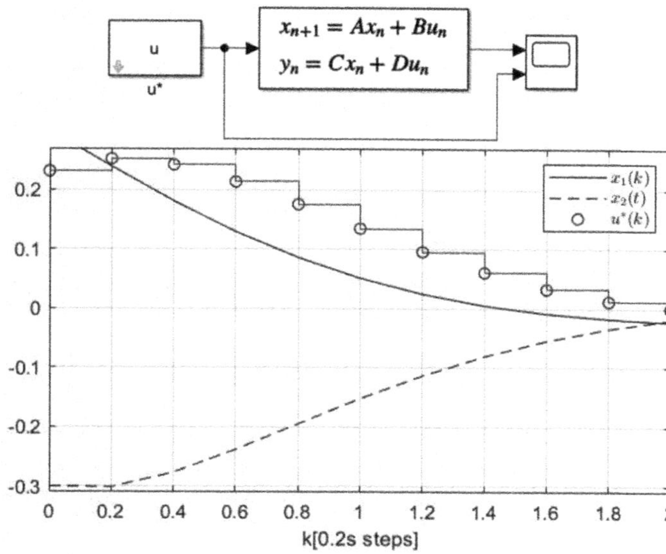

Fig. 9.2 Discrete-time optimal control for plant regulation. The finite horizon of only ten sampling steps is used for demonstration

$$\mathbf{Q} = \begin{bmatrix} 1 & 0 \\ 0 & 1 \end{bmatrix} \quad \text{and} \quad \mathbf{R} = 0.3$$

Then, the matrices $\bar{\mathbf{Q}}$ and $\bar{\mathbf{R}}$ are calculated, referring to (9.4). The number of steps in the control sequence is $N = 2/0.2 = 10$. With N, $\mathbf{A}$, and $\mathbf{B}$ the matrices $\bar{\mathbf{A}}$ and $\bar{\mathbf{B}}$ are determined, referring to (9.6). With $\bar{\mathbf{A}}$, $\bar{\mathbf{Q}}$, $\bar{\mathbf{B}}$, and $\bar{\mathbf{R}}$, the two matrices $\bar{\mathbf{H}}$ and $\bar{\mathbf{F}}$ are determined, referring to (9.7). With $\bar{\mathbf{H}}$, $\bar{\mathbf{F}}$, and $\mathbf{x}(0)$, the optimal open-loop control sequence is determined referring to (9.9). The `Matlab` code in Appendix J with the `Simulink` model shown in Fig. 9.2 simulates the control process, and the result is shown in Fig. 9.2.

9.3 Open-Loop Constrained Finite-Horizon Discrete Linear Optimal Control with Quadratic Cost

When the constraints exist, the optimal control sequence described above will have to be adjusted to satisfy the constraints at all times during the control sequence. The following exercise demonstrates how constraints are included in the optimal control design process.

Exercise

For the plant shown in Sect. 9.2, determine the open-loop ten-point optimal control sequence for 2 s duration with a 0.2 s sampling interval, considering the control signal bounds [–0.2, 0.15].

Solution

When the control signal is constrained, the optimal control problem (9.7) is defined as follows:

$$\min \mathbf{J}(\mathbf{x}(0), \mathbf{u}) = \frac{1}{2}\mathbf{u}^T\bar{\mathbf{H}}\mathbf{u} + \mathbf{x}(0)^T\bar{\mathbf{F}}^T\mathbf{u} + \text{constant}(\mathbf{x}(0))$$

$$\text{subject to } -0.2 \le u(k) \le 0.15\ ;\ k = 1, 2, ...9$$

This is famous quadratic function optimization (Sect. 3.3.1), which involves finding the optimal values for the N elements $\mathbf{u} = u_0, u_1, ...u_{N-1}$ without equality or inequality constraints, but with lower and upper bounds $[u_{min}, u_{max}]$ for each element. Then, the `quadprog` function in `Matlab` can be used as follows:

`>> quadprog(`$\bar{\mathbf{H}}, \bar{\mathbf{F}}\mathbf{x}(0), [\ ], [\ ], [\ ], [\ ], \mathbf{LB}, \mathbf{UB}$) with
$\mathbf{LB} = [-0.2\ \ -0.2\ \ -0.2\ \ -0.2\ \ -0.2\ \ -0.2\ \ -0.2\ \ -0.2\ \ -0.2\ \ -0.2]^T$
$\mathbf{UB} = [0.15\ 0.15\ 0.15\ 0.15\ 0.15\ 0.15\ 0.15\ 0.15\ 0.15\ 0.15]^T$

The `Matlab` code in Appendix K, with the `Simulink` model shown in Fig. 9.2, simulates both unconstrained and constrained control sequences and produces the results shown in Fig. 9.3. The short control horizon $N = 10$ is used to make the `Matlab` code easy to understand. As shown in Fig. 9.3, the optimal control signal drops to nearly zero after $N = 10$ steps, and therefore, the control signals beyond $2s$ are maintained at zero for extended simulation up to $5s$.

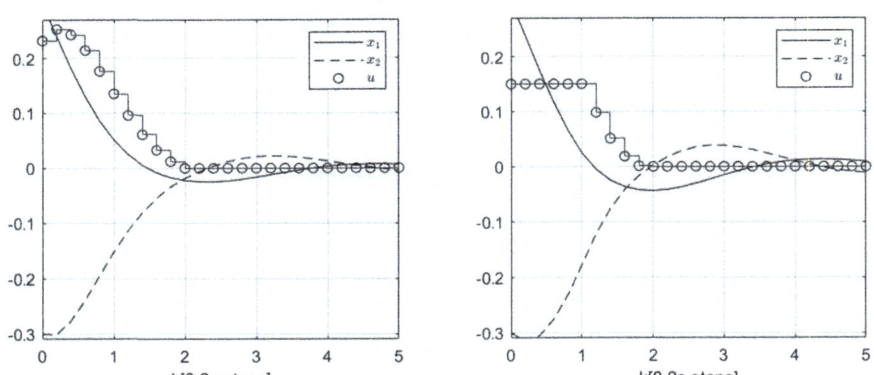

Fig. 9.3 Discrete-time optimal control with and without control bounds for plant regulation. The finite horizon of only ten sampling steps is used, and beyond that, $u = 0$ is used

The optimal control sequence designed offline may not perform well in real implementation due to the presence of unmodeled dynamics and noise that would perturb the plant state in an unpredictable manner, so that the optimal control sequence would become increasingly inappropriate.

The next chapter shows how this problem can be solved by modifying the way the optimal control sequence operates in a real plant with state perturbations. The resulting optimal control process is known as "Model Predictive Control (MPC)," alternatively known as "Receeding Horizon Control."

Chapter 10
Model Predictive Control

The optimal control sequence generated offline does not help realize optimal control in actual systems due to state deviations that are not captured in the state-space model. Such deviations are caused by unmodeled dynamics and noise, leaving only the first few control signals in the optimal control sequence valid and practically useful. In this context, the model predictive control (MPC) is presented in which an optimal control sequence is generated for a finite time into the future (the prediction horizon), and only the first few control signals (the control horizon) are practically used. After each control horizon, the process recurs in its actual state at that time.

10.1 Receeding Horizon

Figure 10.1 shows a $N_p = 4$ prediction horizon and $N_c = 1$ control horizon. In this format, at each step, the next four optimal control signals are calculated, but only the first one of the four signals is applied. The process recurs at each step based on the actual state observed.

10.2 Model Predictive Regulator

This chapter presents the model predictive regulator with unconstrained and constrained control.

S. R. Munasinghe, *Optimization and Optimal Control in a Nutshell*, Engineering Optimization: Methods and Applications,
https://doi.org/10.1007/978-981-97-8167-6_10

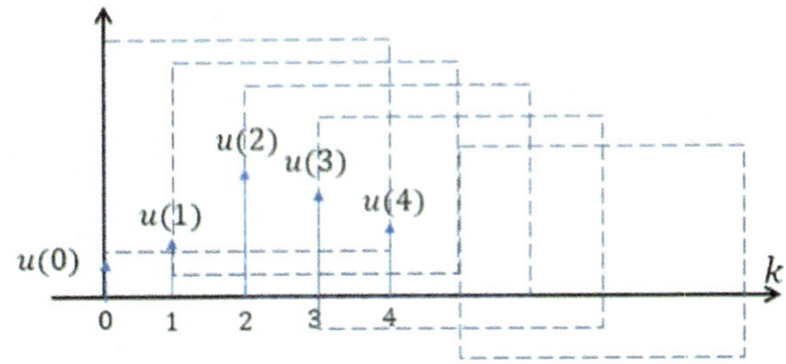

Fig. 10.1 Model predictive control with a four-step prediction horizon and a one-step control horizon. Dotted rectangles show the receding horizon

10.2.1 Unconstrained Model Predictive Regulator

Referring to (9.9), at time k, the control sequence for future N_p steps is determined as follows:

$$\mathbf{u}^*(k) = -\bar{\mathbf{H}}^{-1}\bar{\mathbf{F}}\mathbf{x}(k) \tag{10.1}$$

Then, the first control signal of this sequence is $u_1^*(k)$. The first control signal of this sequence, $u_1^*(k)$, is picked up as follows:

$$\begin{aligned}\mathbf{u}_1^*(k) &= -[1 \quad 0 \quad 0 \quad 0...0]\bar{\mathbf{H}}^{-1}\bar{\mathbf{F}}\mathbf{x}(k) \\ &= -K\mathbf{x}(k)\end{aligned} \tag{10.2}$$

where $K = -[1 \quad 0 \quad 0 \quad 0...0]\bar{\mathbf{H}}^{-1}\bar{\mathbf{F}}$. This reveals that the model predictive regulator uses a type of state feedback control. The model predictive regulator algorithm is shown below, where the state-space model $\mathbf{A}$, $\mathbf{B}$, cost weights $\mathbf{Q}$, $\mathbf{R}$, control duration N, and prediction horizon N_p are given beforehand.

The matrices $\bar{\mathbf{Q}}$, $\bar{\mathbf{R}}$, $\bar{\mathbf{A}}$, $\bar{\mathbf{B}}$, $\bar{\mathbf{H}}$, and $\bar{\mathbf{F}}$ are calculated using (9.4), (9.6), and (9.7). The control loop is iterated $N - N_p$ times, and in each iteration, the optimal control sequence for future N_p steps (prediction horizon) is calculated based on the present state $\mathbf{x}(k)$. And, in each iteration, the first control signal is picked up and used to actuate the plant.

10.2.2 Constrained Model Predictive Regulator

When the constraints exist, the model predictive regulator optimizes the control signal sequence without violating the constraints. The following algorithm presents how the `quadprog` command in `Matlab` is used to implement constraints in the model predictive regulator.

10.2.3 *Model Predictive Regulator Design*

Exercise:

Develop the finite-time optimal regulator in Exercise 9.3 into a model predictive regulator considering unbound controls. Introduce control bounds and implement the model predictive regulator using `quadprog` optimization in `Matlab`.

Solution:

The unconstrained and constrained model predictive regulator algorithms presented in Figs. 10.2 and 10.3 were used on the plant presented in Sect. 9.3 with a control duration of 6*s* and a prediction horizon of 10 steps. Control was constrained to the

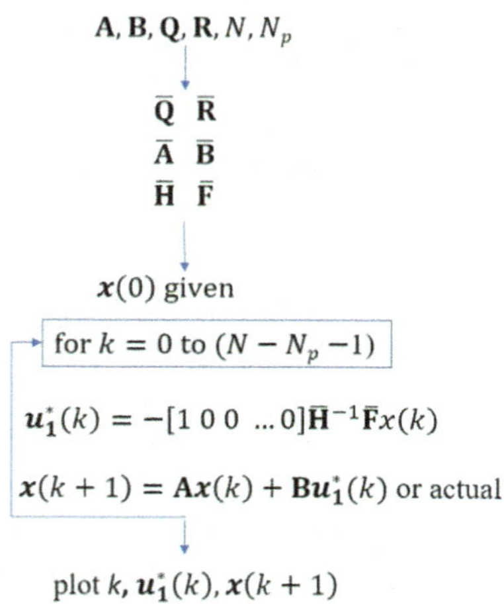

Fig. 10.2 Unconstrained model predictive regulator algorithm

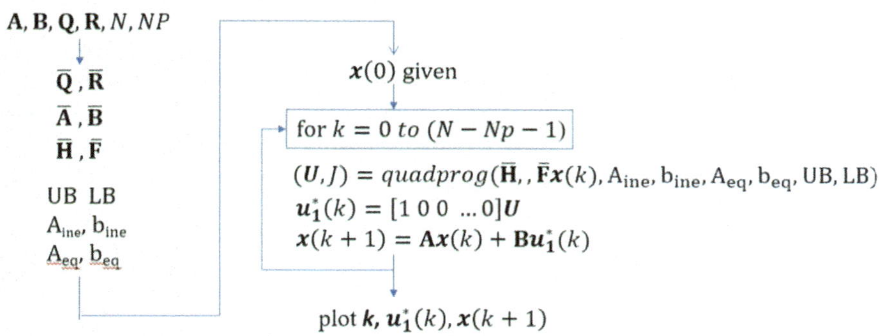

Fig. 10.3 Constrained model predictive regulator algorithm

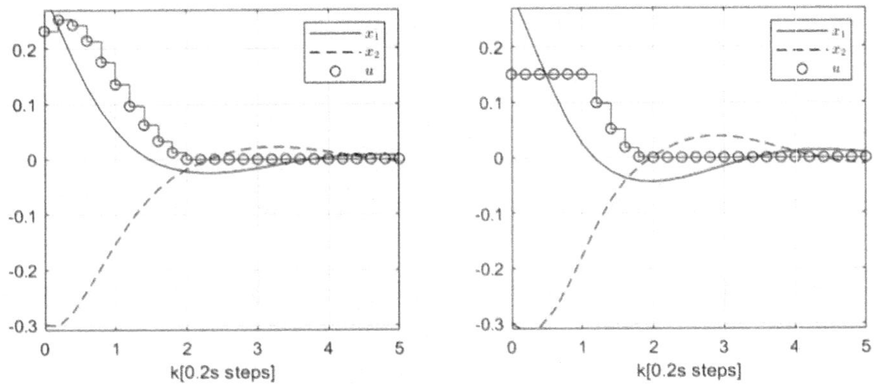

Fig. 10.4 Discrete-time model predictive regulator with unconstrained and constrained control

range [–0.2, 0.15], and the sampling interval was set to 0.2 s. The `Matlab` code of the exercise is available in Appendix L. Regulator performance with unbound and bounded controls is shown in Fig. 10.4.

10.3 Model Predictive Tracking Control

Unlike in regulator design, tracking control does not target the zero state at the end of the predicted horizon; therefore, a sustainable non-zero control signal is required at the end state. This requirement is somewhat contradictory to the optimal control policy, where the control signal is also reduced to minimize cost. Therefore, the $p \times 1$ control signal vector is written as follows:

$$\mathbf{u}(k) = \mathbf{u}(k-1) + \Delta\mathbf{u}(k) \quad ; \quad k = 0, 1, 2...N_p - 1, \mathbf{u}(-\mathbf{1}) = \mathbf{0} \tag{10.3}$$

and the minimization is applied to the incremental component $\Delta\mathbf{u}(k)$ only. Hence, the optimization process is applied to the cost associated with $\Delta\mathbf{u} = [\Delta\mathbf{u}(0)\ \Delta\mathbf{u}(1)\ ...\ \Delta\mathbf{u}(N_p-1)]$, instead of the cost of the control sequence $\mathbf{u} = [\mathbf{u}(0)\ \mathbf{u}(1)\ ...\ \mathbf{u}(N_p-1)]$. With the redefinition of the control signal in (10.3), the state-space model is as follows:

$$\begin{aligned}\mathbf{x}(k+1) &= \mathbf{A}\mathbf{x}(k) + \mathbf{B}\mathbf{u}(k)\\ &= \mathbf{A}\mathbf{x}(k) + \mathbf{B}\mathbf{u}(k-1) + \mathbf{B}\Delta\mathbf{u}(k)\\ &= \left[\mathbf{A}\ \mathbf{B}\right]\begin{bmatrix}\mathbf{x}(k)\\ \mathbf{u}(k-1)\end{bmatrix} + \mathbf{B}\Delta\mathbf{u}(k)\end{aligned} \tag{10.4}$$

Let's combine (10.3) and (10.4) to generate the augmented state-space model as follows:

$$\begin{bmatrix}\mathbf{x}(k+1)\\ \mathbf{u}(k)\end{bmatrix}=\begin{bmatrix}\mathbf{A} & \mathbf{B}\\ 0 & I\end{bmatrix}\begin{bmatrix}\mathbf{x}(k)\\ \mathbf{u}(k-1)\end{bmatrix}+\begin{bmatrix}\mathbf{B}\\ I\end{bmatrix}\Delta\mathbf{u}(k)$$
$$\tilde{\mathbf{x}}(k+1)=\tilde{\mathbf{A}}\tilde{\mathbf{x}}(k)+\tilde{\mathbf{B}}\Delta\mathbf{u}(k) \tag{10.5}$$

where $\tilde{\mathbf{A}}$ is a $(n+m)\times(n+m)$ matrix and $\tilde{\mathbf{B}}$ is a $(n+m)\times m$ matrix for a n state and m controls. Accordingly, the $p\times 1$ output vector is written as follows:

$$\begin{aligned}\mathbf{y}(k)&=\begin{bmatrix}\mathbf{C} & 0\end{bmatrix}\tilde{\mathbf{x}}(k)\\ &=\tilde{\mathbf{C}}\tilde{\mathbf{x}}(k)\end{aligned} \tag{10.6}$$

where $\tilde{\mathbf{C}}$ is a $p\times(n+m)$ matrix. Then, the $p\times 1$ reference vector $\mathbf{r}(k)$ and tracking error vector $\mathbf{e}(k)$ are defined as follows:

$$\begin{aligned}\mathbf{e}(k)&=\mathbf{r}(k)-\mathbf{y}(k)\quad ;\quad k=1,2,...N\\ &=\mathbf{r}(k)-\tilde{\mathbf{C}}\tilde{\mathbf{x}}(k)\end{aligned} \tag{10.7}$$

Here, $\mathbf{e}(1)$ is the first error term because it is the error term associated with the first control $\mathbf{u}(0)$. Figure 10.5 shows the N_p-step prediction horizon at time t, the augmented state sequence $\tilde{\mathbf{x}}(k)$, the incremental control sequence $\Delta\mathbf{u}(k)$, and the reference sequence $\mathbf{r}$(k).

With the differential control signal and the tracking error, the cost of optimal tracking control over the prediction horizon starting from an arbitrary time t is defined as follows:

$$\begin{aligned}\mathbf{J}=&\frac{1}{2}\mathbf{e}(N_p)^T\mathbf{Q}_{N_p}\mathbf{e}(N_p)+\frac{1}{2}\sum_{k=1}^{N_p-1}\mathbf{e}(k)^T\mathbf{Q}\mathbf{e}(k)+\frac{1}{2}\sum_{k=0}^{N_p-1}\Delta\mathbf{u}(k)^T\mathbf{R}\Delta\mathbf{u}(k)\\ =&\frac{1}{2}\left[\mathbf{r}(N_p)-\tilde{\mathbf{C}}\tilde{\mathbf{x}}(N_p)\right]^T\mathbf{Q}_{N_p}\left[\mathbf{r}(N_p)-\tilde{\mathbf{C}}\tilde{\mathbf{x}}(N_p)\right]+\\ &\frac{1}{2}\sum_{k=1}^{N_p-1}\left[\mathbf{r}(k)-\tilde{\mathbf{C}}\tilde{\mathbf{x}}(k)\right]^T\mathbf{Q}\left[\mathbf{r}(k)-\tilde{\mathbf{C}}\tilde{\mathbf{x}}(k)\right]+\frac{1}{2}\sum_{k=0}^{N_p-1}\Delta\mathbf{u}(k)^T\mathbf{R}\Delta\mathbf{u}(k)\end{aligned} \tag{10.8}$$

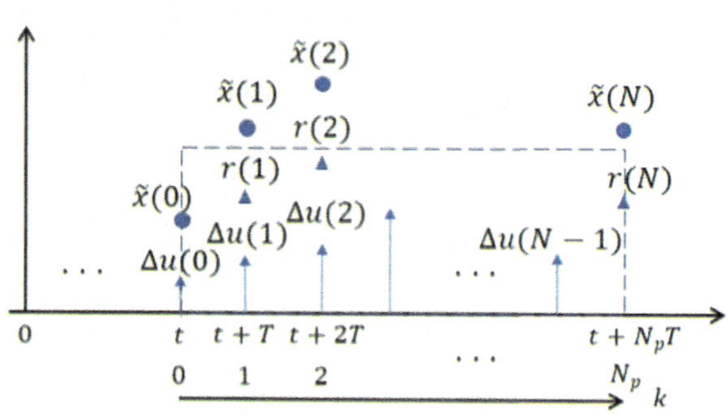

Fig. 10.5 The N_p-step prediction horizon at time t of a model predictive single state tracking control

where $\mathbf{Q}_{N_p}$, $\mathbf{Q}$, and $\mathbf{R}$ are cost weight matrices of the final state error, tracking state error, and the incremental control effort. The first term on the right-hand side is the cost due to the tracking error $\mathbf{e}(N_p)$ at the end of the prediction horizon; the second term is the running cost of the tracking error $\mathbf{e}(k)$; and the third term is the running cost of the incremental control signal $\Delta\mathbf{u}(k)$. The quadratic cost can be expanded as follows:

$$
\begin{aligned}
\mathbf{J} =& \frac{1}{2}\mathbf{r}(N_p)^T\mathbf{Q}_{N_p}\mathbf{r}(N_p) - \mathbf{r}(N_p)^T\mathbf{Q}_{N_p}\tilde{\mathbf{C}}\tilde{\mathbf{x}}(N_p) + \frac{1}{2}\tilde{\mathbf{x}}(N_p)^T\tilde{\mathbf{C}}^T\mathbf{Q}_{N_p}\tilde{\mathbf{C}}\tilde{\mathbf{x}}(N_p) + \\
& \frac{1}{2}\sum_{k=1}^{N_p-1}\left\{\mathbf{r}(k)^T\mathbf{Q}\mathbf{r}(k) - 2\mathbf{r}(k)^T\mathbf{Q}\tilde{\mathbf{C}}\tilde{\mathbf{x}}(k) + \tilde{\mathbf{x}}(k)^T\tilde{\mathbf{C}}^T\mathbf{Q}\tilde{\mathbf{C}}\tilde{\mathbf{x}}(k)\right\} + \\
& \frac{1}{2}\sum_{k=0}^{N_p-1}\Delta\mathbf{u}(k)^T\mathbf{R}\Delta\mathbf{u}(k)
\end{aligned} \tag{10.9}
$$

The first and fourth terms solely depend on the reference trajectory; hence, they do not play a part in the optimization process. Once these two terms are dropped, and combining the second and fifth terms and the third and sixth terms, the cost can be written in the following matrix form:

$$
\begin{aligned}
\mathbf{J} =& \frac{1}{2}\tilde{\mathbf{x}}^T\begin{bmatrix} \tilde{\mathbf{C}}^T\mathbf{Q}\tilde{\mathbf{C}} & 0 & 0 & \dots & 0 \\ 0 & \tilde{\mathbf{C}}^T\mathbf{Q}\tilde{\mathbf{C}} & 0 & \dots & 0 \\ 0 & 0 & \tilde{\mathbf{C}}^T\mathbf{Q}\tilde{\mathbf{C}} & \dots & 0 \\ & & & \vdots & \\ 0 & 0 & 0 & \dots & \tilde{\mathbf{C}}^T\mathbf{Q}_N\tilde{\mathbf{C}} \end{bmatrix}\tilde{\mathbf{x}} \\
& - \mathbf{r}^T\begin{bmatrix} \mathbf{Q}\tilde{\mathbf{C}} & 0 & 0 & \dots & 0 \\ 0 & \mathbf{Q}\tilde{\mathbf{C}} & 0 & \dots & 0 \\ 0 & 0 & \mathbf{Q}\tilde{\mathbf{C}} & \dots & 0 \\ & & & \vdots & \\ 0 & 0 & 0 & \dots & \mathbf{Q}_N\tilde{\mathbf{C}} \end{bmatrix}\tilde{\mathbf{x}} \\
& + \Delta\mathbf{u}^T\begin{bmatrix} \mathbf{R} & 0 & 0 & \dots & 0 \\ 0 & \mathbf{R} & 0 & \dots & 0 \\ & & & \vdots & \\ 0 & 0 & 0 & \dots & \mathbf{R} \end{bmatrix}\Delta\mathbf{u} \\
=& \frac{1}{2}\tilde{\mathbf{x}}^T\bar{\bar{\mathbf{Q}}}\tilde{\mathbf{x}} - \mathbf{r}^T\bar{\bar{\mathbf{T}}}\tilde{\mathbf{x}} + \frac{1}{2}\Delta\mathbf{u}^T\bar{\bar{\mathbf{R}}}\Delta\mathbf{u}
\end{aligned} \tag{10.10}
$$

where $\mathbf{r} = [\mathbf{r}(1)\ \mathbf{r}(2)\ \dots\ \mathbf{r}(N_p)]^T$ is a $pN_p \times 1$ vector, $\tilde{\mathbf{x}} = [\tilde{\mathbf{x}}(1)\ \tilde{\mathbf{x}}(2)\ \dots\ \tilde{\mathbf{x}}(N_p)]^T$ is a $(n+m)N_p \times 1$ vector, and $\Delta\mathbf{u} = [\Delta\mathbf{u}(0)\ \Delta\mathbf{u}(1)\ \dots\ \Delta\mathbf{u}(N_p-1)]^T$ is a $mN_p \times 1$

vector. Accordingly, $\bar{\bar{\mathbf{Q}}}$ is a $(n+m)N_p \times (n+m)N_p$ matrix, $\bar{\bar{\mathbf{T}}}$ is a $pN_p \times (n+m)N_p$ matrix, and $\bar{\bar{\mathbf{R}}}$ is a $mN_p \times mN_p$ matrix.

Recalling from (9.6) $\tilde{\mathbf{x}}$ can be eliminated using the state transition as follows:

$$\begin{bmatrix} \tilde{\mathbf{x}}(1) \\ \tilde{\mathbf{x}}(2) \\ \tilde{\mathbf{x}}(3) \\ \vdots \\ \tilde{\mathbf{x}}(N) \end{bmatrix} = \begin{bmatrix} \tilde{\mathbf{A}} \\ \tilde{\mathbf{A}}^2 \\ \tilde{\mathbf{A}}^3 \\ \vdots \\ \tilde{\mathbf{A}}^N \end{bmatrix} \tilde{\mathbf{x}}(0) + \begin{bmatrix} \tilde{\mathbf{B}} & 0 & 0 & \dots 0 \\ \tilde{\mathbf{A}}\tilde{\mathbf{B}} & \tilde{\mathbf{B}} & 0 & \dots 0 \\ \tilde{\mathbf{A}}^2\tilde{\mathbf{B}} & \tilde{\mathbf{A}}\tilde{\mathbf{B}} & \tilde{\mathbf{B}} & \dots 0 \\ & & & \vdots \\ \tilde{\mathbf{A}}^{N-1}\tilde{\mathbf{B}} & \tilde{\mathbf{A}}^{N-2}\tilde{\mathbf{B}} & \tilde{\mathbf{A}}^{N-3}\tilde{\mathbf{B}} & \dots \tilde{\mathbf{B}} \end{bmatrix} \begin{bmatrix} \Delta\mathbf{u}(0) \\ \Delta\mathbf{u}(1) \\ \Delta\mathbf{u}(2) \\ \vdots \\ \Delta\mathbf{u}(N-1) \end{bmatrix}$$

$$\tilde{\mathbf{x}} = \bar{\bar{\mathbf{A}}}\tilde{\mathbf{x}}(0) + \bar{\bar{\mathbf{B}}}\Delta\mathbf{u} \tag{10.11}$$

where $\bar{\bar{\mathbf{A}}}$ is a $(n+m)N_p \times (n+m)$ matrix, and $\bar{\bar{\mathbf{B}}}$ is a $(n+m)N_p \times mN_p$ matrix. Then, the cost (10.10) can be written by substituting for $\tilde{\mathbf{x}}$ from (10.11) as follows:

$$\begin{aligned} \mathbf{J} =& \frac{1}{2}\left\{(\bar{\bar{\mathbf{A}}}\tilde{\mathbf{x}}(0) + \bar{\bar{\mathbf{B}}}\Delta\mathbf{u})^T\bar{\bar{\mathbf{Q}}}(\bar{\bar{\mathbf{A}}}\tilde{\mathbf{x}}(0) + \bar{\bar{\mathbf{B}}}\Delta\mathbf{u})\right\} - \mathbf{r}^T\bar{\bar{\mathbf{T}}}(\bar{\bar{\mathbf{A}}}\tilde{\mathbf{x}}(0) + \bar{\bar{\mathbf{B}}}\Delta\mathbf{u}) + \frac{1}{2}\Delta\mathbf{u}^T\bar{\bar{\mathbf{R}}}\Delta\mathbf{u} \\ =& \frac{1}{2}\left\{\Delta\mathbf{u}^T\bar{\bar{\mathbf{B}}}^T\bar{\bar{\mathbf{Q}}}\bar{\bar{\mathbf{B}}}\Delta\mathbf{u} + 2\tilde{\mathbf{x}}(0)^T\bar{\bar{\mathbf{A}}}^T\bar{\bar{\mathbf{Q}}}\bar{\bar{\mathbf{B}}}\Delta\mathbf{u} + \tilde{\mathbf{x}}(0)^T\bar{\bar{\mathbf{A}}}^T\bar{\bar{\mathbf{Q}}}\bar{\bar{\mathbf{A}}}\tilde{\mathbf{x}}(0)\right\} - \mathbf{r}^T\bar{\bar{\mathbf{T}}}\bar{\bar{\mathbf{A}}}\tilde{\mathbf{x}}(0) \\ & - \mathbf{r}^T\bar{\bar{\mathbf{T}}}\bar{\bar{\mathbf{B}}}\Delta\mathbf{u} + \frac{1}{2}\Delta\mathbf{u}^T\bar{\bar{\mathbf{R}}}\Delta\mathbf{u} \end{aligned} \tag{10.12}$$

where the third and fourth terms are constants that depend solely on the initial state $\mathbf{x}(0)$ and the reference trajectory $\mathbf{r}$, hence non-contributing to the optimization. Once these two terms are dropped, the cost takes the following form:

$$\begin{aligned} \mathbf{J} =& \frac{1}{2}\Delta\mathbf{u}^T\bar{\bar{\mathbf{B}}}^T\bar{\bar{\mathbf{Q}}}\bar{\bar{\mathbf{B}}}\Delta\mathbf{u} + \tilde{\mathbf{x}}(0)^T\bar{\bar{\mathbf{A}}}^T\bar{\bar{\mathbf{Q}}}\bar{\bar{\mathbf{B}}}\Delta\mathbf{u} - \mathbf{r}^T\bar{\bar{\mathbf{T}}}\bar{\bar{\mathbf{B}}}\Delta\mathbf{u} + \frac{1}{2}\Delta\mathbf{u}^T\bar{\bar{\mathbf{R}}}\Delta\mathbf{u} \\ =& \frac{1}{2}\Delta\mathbf{u}^T(\bar{\bar{\mathbf{B}}}^T\bar{\bar{\mathbf{Q}}}\bar{\bar{\mathbf{B}}} + \bar{\bar{\mathbf{R}}})\Delta\mathbf{u} + \left[\tilde{\mathbf{x}}(0)^T \ \mathbf{r}^T\right]\begin{bmatrix} \bar{\bar{\mathbf{A}}}^T\bar{\bar{\mathbf{Q}}}\bar{\bar{\mathbf{B}}} \\ -\bar{\bar{\mathbf{T}}}\bar{\bar{\mathbf{B}}} \end{bmatrix}\Delta\mathbf{u} \\ =& \frac{1}{2}\Delta\mathbf{u}^T\bar{\bar{\mathbf{H}}}\Delta\mathbf{u} + \left[\tilde{\mathbf{x}}(0)^T \ \mathbf{r}^T\right]\bar{\bar{\mathbf{F}}}^T\Delta\mathbf{u} \end{aligned} \tag{10.13}$$

where $\bar{\bar{\mathbf{H}}}$ is a $mN_p \times mN_p$ matrix and $\bar{\bar{\mathbf{F}}}$ is a $(n+m+pN_p) \times mN_p$ matrix. The cost $\mathbf{J}$ over the N_p-step prediction horizon depends on the N_p incremental control signals $\Delta\mathbf{u}$ and the augmented initial state.

$$\tilde{\mathbf{x}}(0) = \begin{bmatrix} \mathbf{x}(0) \\ \mathbf{u}(-1) \end{bmatrix}$$

By differentiating the cost with respect to the N_p incremental control signals and setting it to zero, the FONC for optimal control is determined as follows:

$$\nabla_{\Delta \mathbf{u}^*} \mathbf{J} = \bar{\bar{\mathbf{H}}} \Delta \mathbf{u}^* + \bar{\bar{\mathbf{F}}} \begin{bmatrix} \tilde{\mathbf{x}}(0) \\ \mathbf{r} \end{bmatrix} = 0 \tag{10.14}$$

Then, the optimal incremental control sequence for tracking a given reference trajectory starting from a given initial state is as follows:

$$\Delta \mathbf{u}^* = -\bar{\bar{\mathbf{H}}}^{-1} \bar{\bar{\mathbf{F}}} \begin{bmatrix} \tilde{\mathbf{x}}(0) \\ \mathbf{r} \end{bmatrix} \tag{10.15}$$

The model predictive tracking controller at each step $k = 0,\ 1,\ 2,\ 3 \ldots N$ determines the future N_p differential control signals as follows:

$$\Delta \mathbf{u}^*(k) = -\bar{\bar{\mathbf{H}}}^{-1} \bar{\bar{\mathbf{F}}} \begin{bmatrix} \tilde{\mathbf{x}}(k) \\ r(k) \\ r(k+1) \\ \vdots \\ r(k+N_p-1) \end{bmatrix} \tag{10.16}$$

Referring to (10.16) and Fig. 10.5 the reference trajectory $\mathbf{r}$ needs to be planned with N_p additional steps beyond the actual control duration N. The model predictive tracking controller algorithm is shown in Fig. 10.6.

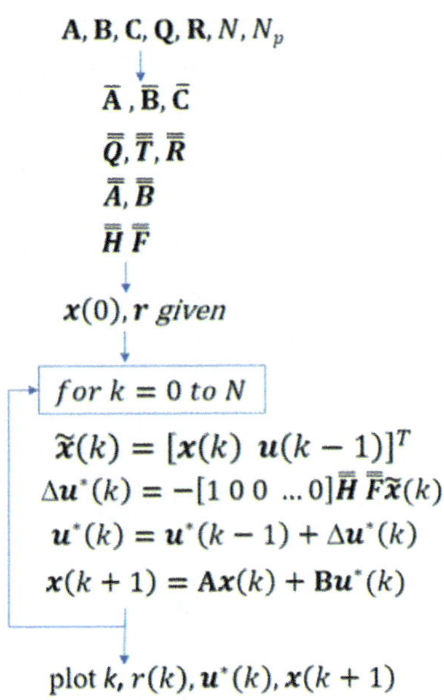

Fig. 10.6 Model predictive tracking control algorithm

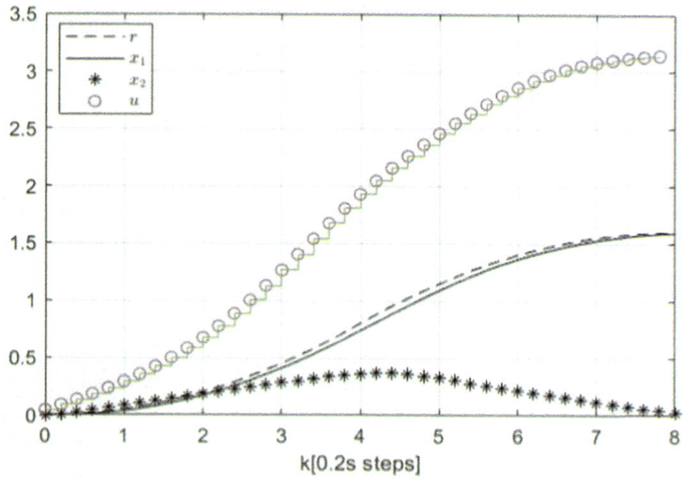

Fig. 10.7 Unconstrained model predictive tracking control result

10.3.1 Unconstrained Model Predictive Tracking

Exercise:

The example in 10.2.3 is modified into a tracking control problem as follows: The first state variable $x_1(k)$ is to follow a reference trajectory $r(k)$ for which a model predictive tracking controller is required. The reference trajectory is derived assuming a uniform acceleration of 0.1 in the first 4 s and a uniform deceleration of –0.1 in the latter 4 s of the 8 s control duration. The Matlab code of the tracking controller is given in Appendix M. The result of tracking control together with the reference trajectory is shown in Fig. 10.7.

10.3.2 Constrained Model Predictive Tracking

Exercise:

The exercise in 10.3.1 is repeated here with the limits on the control signal. The `quadprog` command is used to determine the feasible optimal control signal $\mathbf{u}(k)$ at time k subject to the upper and lower bounds. At time k, the prediction horizon is $[k, k + N_p]$, hence the `quadprog` command generates an optimal control sequence $u(k), u(k+1), ...u(k+N_p-1)$ from which only the first optimal control signal $\mathbf{u}(k)$ is used for action. Therefore, the upper and lower bounds are only applied to the first control signal $\mathbf{u}(k)$, and the other $N_p - 1$ control signals are considered unbounded. For example, a five-step prediction horizon has the following vectors for upper and lower bounds:

$$UB = [U_{max} \;\; +\infty \;\; +\infty \;\; +\infty \;\; +\infty]^T$$
$$LB = [U_{min} \;\; -\infty \;\; -\infty \;\; -\infty \;\; -\infty\,]^T$$

However, in tracking control, the optimization is formulated on the incremental control signal $\Delta u(k)$ as given in (10.5); hence, at each k, the incremental control signal is to be constrained as follows:

$$UB = [U_{max} - u(k-1) \;\; +\infty \;\; +\infty \;\; +\infty \;\; +\infty\,]$$
$$LB = [U_{min} - u(k-1) \;\; -\infty \;\; -\infty \;\; -\infty \;\; -\infty]$$

so that $u(k) = u(k-1) + \Delta u(k)$ is maintained within $[LB, UB]$ at time step k. The control algorithm for the constrained model predictive tracking control is shown in Fig. 10.8, and the `Matlab` code is given in Appendix N.

The tracking control results with and without control signal bounds are shown in Fig. 10.9, where it is shown that the control signal hits the upper bound and stays saturated, causing a growing tracking error. It is noteworthy to state that the optimal control cannot guarantee accurate tracking performance if the reference trajectory causes actuator saturation. Hence, it is necessary to plan a feasible reference so that tracking accuracy and optimal performance can be achieved at the same time.

Summary

The field of model predictive control was started in 1980 with the publication of the seminal paper (Cutler and Ramaker 1980) on industrial process control. And today, model predictive control is heavily used in many fields due to its simplicity

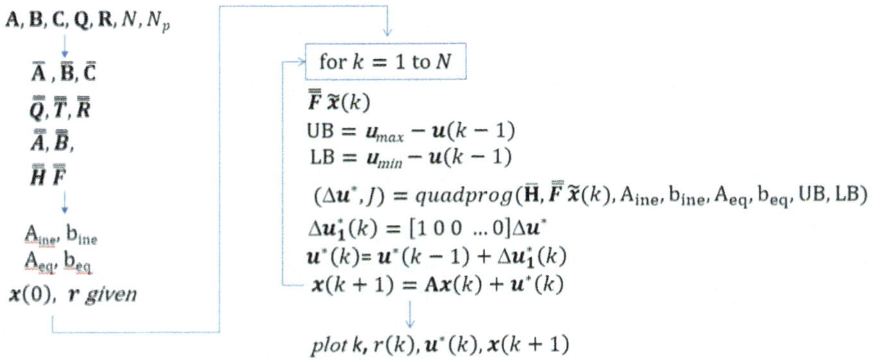

Fig. 10.8 Constrained model predictive tracking control algorithm

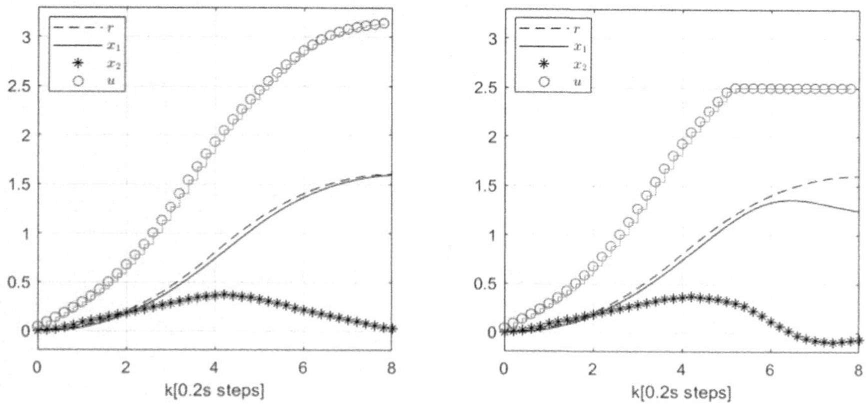

Fig. 10.9 Performance deterioration when the control signal is constrained

in computer-based discrete implementation and its ongoing optimization with state updates. A comprehensive account of the development of the field of model predictive control is presented in Schwenzer et al. (2021).

References

Cutler CR, Ramaker BL (1980) Dynamic matrix control—a computer control algorithm. In: Joint automatic control conference, vol 1, WP5-B, San Francisco, CA

Schwenzer M, Ay M, Bergs T, Abel D (2021) Review on model predictive control—an engineering perspective. Int J Adv Manuf Technol 1327–1349

Appendix A
Preliminaries

An increment of a function is defined as $\delta f = f(t + \delta t) - f(t)$ (A.1)

A differential of a function is defined as $\frac{df(t)}{dt} = \lim_{\delta t \to 0} \frac{\delta f(t)}{\delta t}$ (A.2)

A functional is defined as $J(f(x)) = \int_{x_1}^{x_2} f(x)dx$ (A.3)

First variation and second variation of a functional are defined as $\frac{\partial}{\partial x} J(f(x))$ and $\frac{\partial^2}{\partial x^2} J(f(x))$, respectively. An increment of a functional is defined as

$$\delta J = J(x(t) + \delta x(t)) - J(x(t)) \tag{A.4}$$

$$= \frac{\partial}{\partial x} J(f(x)) + \frac{1}{2!} \frac{\partial^2}{\partial x^2} J(f(x)) + HoT \tag{A.5}$$

Some identities of Vector Calculus

$$\frac{\partial(A\mathbf{x})}{\partial \mathbf{x}} = A \tag{A.6}$$

$$\frac{\partial(\mathbf{x}^T \mathbf{y})}{\partial \mathbf{y}} = \frac{\partial(\mathbf{y}^T \mathbf{x})}{\partial \mathbf{y}} = \mathbf{x} \tag{A.7}$$

$$\frac{\partial(\mathbf{x}^T A \mathbf{y})}{\partial \mathbf{y}} = \frac{\partial(\mathbf{y}^T A^T \mathbf{x})}{\partial \mathbf{y}} = A^T \mathbf{x} \tag{A.8}$$

S. R. Munasinghe, *Optimization and Optimal Control in a Nutshell*, Engineering Optimization: Methods and Applications,
https://doi.org/10.1007/978-981-97-8167-6

$$\frac{\partial(\mathbf{x}^T A\mathbf{x})}{\partial \mathbf{x}} = A\mathbf{x} + A^T\mathbf{x} \tag{A.9}$$

$$\begin{aligned}\frac{\partial^2(\mathbf{x}^T A\mathbf{x})}{\partial \mathbf{x}^2} &= A + A^T \\ &= 2A \text{ If } A \text{ is Symmetric}\end{aligned} \tag{A.10}$$

Leibniz Integral Rule

$$\frac{d}{dx}\int_{a(x)}^{b(x)} f(t,x)dt = \int_{a(x)}^{b(x)} \frac{\partial}{\partial x} f(t,x)dt \tag{A.11}$$

Appendix B
Matlab Code 2.3.1

```
% Unconstrained Quadratic Function Optimization

clear; close all;
[x1,x2] = meshgrid(-5:0.4:5,-5:0.4:5);

Q = [2 0;
     0 4];
r = [2;
    -1];

cost = 0.5*(Q(1,1)*x1.^2 + Q(2,2)*x2.^2) + r(1)*x1 +r(2)*x2 + 3;

subplot(121); contour(x1,x2,cost,20,'LineWidth',2); hold on;
subplot(122); surf(x1,x2,cost); hold on;

A=[];b=[];Aeq=[];beq=[];LB=[];UB=[]; % No constraints
[X,J] = quadprog(Q,r,A,b,Aeq,beq,LB,UB) % Unconstrained optimization
subplot(121); plot(X(1),X(2),'r.','MarkerSize',20);
xlabel('x1'); ylabel('x2'); title('Contour lines')
subplot(122); plot3(X(1),X(2),J+3,'r.','MarkerSize',20);
xlabel('x1'); ylabel('x2'); zlabel('cost'); title('Cost Surface')
```


S. R. Munasinghe, *Optimization and Optimal Control in a Nutshell*, Engineering Optimization: Methods and Applications,
https://doi.org/10.1007/978-981-97-8167-6

Appendix C
Matlab Code 3.3.1

```
% Quadratic Optimization with two linear inequality constraints
clear; close all;
[x1,x2]=meshgrid(0:0.2:10,0:0.2:10); % creates 2D data point cloud

Q = [0.8 0;
     0  2];
r = [-5;
     -6];
c = 50;

% Quadratic Cost 0.4x1^2 + x2^2 -5x1 -6x2 + 50
cost = 0.5*(Q(1,1)*x1.^2 + Q(2,2)*x2.^2) + r(1)*x1 + r(2)*x2 + c;

% Plot 2D contour map and constraints
subplot(121); contour(x1,x2,cost,20,'LineWidth',2); hold on;
xlabel('x_1'); ylabel('x_2');

% Plot 3D Surface and constraints on 2D domain
subplot(122);surf(x1,x2,cost); hold on;
xlabel('x_1'); ylabel('x_2'); zlabel('Cost');hold on;

% Linear Inequality constraints
% x1-x2 < -2 AND -0.3x1-x2 < -8
A = [1    -1;
    -0.3  -1];
b = [-2;
     -8];

% No linear equality constraints
Aeq = []; beq = [];

% Limits of State variables
LB = [0;1];
UB = [10;8];
subplot(121); axis([LB(1) UB(1) LB(2) UB(2)]);
plot([0;8],[2;10],'k','LineWidth',2);
plot([0;10],[8;5],'k','LineWidth',2);

subplot(122); axis([LB(1) UB(1) LB(2) UB(2) 0 70]);
plot([0;8],[2;10],'k','LineWidth',2);
plot([0;10],[8;5],'k','LineWidth',2);

% Run quadratic Program
[X,J] = quadprog(Q,r,A,b,Aeq,beq,LB,UB)
```


S. R. Munasinghe, *Optimization and Optimal Control in a Nutshell*, Engineering Optimization: Methods and Applications,
https://doi.org/10.1007/978-981-97-8167-6

```
% Plot the minima on the 2D contour
subplot(121); plot(X(1),X(2),'r.','MarkerSize',25);
title('Cost Contours')

% Plot the minima on the 3D surface
mincost = 0.5*(Q(1,1)*X(1)^2 + Q(2,2)*X(2)^2) +...
    r(1)*X(1) + r(2)*X(2) + c;
subplot(122); plot3(X(1),X(2),mincost,'r.','MarkerSize',25);
title('Cost Surface')
```

Appendix D
Matlab Code 3.3.2

```
% Arbitrary Function Optimization
% subject to Nonlinear Equality Constraint
clear; close all;

% cost surface
[x1,x2] = meshgrid(-5:0.1:1, -3:0.1:3);
cost = sin(x1)+0.05*x1.^2+0.1*x2.^2;
subplot(121); contour(x1,x2,cost,20,'LineWidth',2); hold on;
title('Cost Contours')
xlabel('x1'); ylabel('x2');
subplot(122); surf(x1,x2,cost); hold on;
title('Cost Surface');
xlabel('x1'); ylabel('x2'); zlabel('Cost');

% constraints
LB = [-5;-3]; UB = [1;3]; % limits of state varuables
x0 = [-3;-3];             % initial solution
A = [1 1]; b = [3];       % LInear equality constriants
Aeq = []; beq = [];       % No linear constraints

%fmincon constrained optimization
[xstar, fstar] = fmincon(@calcost,x0,A,b,Aeq,beq,LB,UB,@NLconstraints,[])

plot3(xstar(1),xstar(2),fstar,'r.','MarkerSize',20); hold on;

%Plot constraint
x = -5:0.1:1;
ynl = (x+3).^3; % nonlinear equality constraint
yl = -x-0.5; %linear inequality constraint

subplot(121);
plot(x,ynl,'k','LineWidth',2); axis([-5 1 -3 3]);
plot(x,yl,'k','LineWidth',2); axis([-5 1 -3 3]);
plot(xstar(1),xstar(2),'r.','MarkerSize',20);

subplot(122);
plot(x,ynl,'k','LineWidth',2); axis([-5 1 -3 3 -1 3.5]);

function cost = calcost(x)
cost = sin(x(1))+0.05*x(1)^2+0.1*x(2)^2;
end

function [c,ceq]=NLconstraints(x)
ceq=(x(1)+3)^3-x(2);
c = [];
end
```


S. R. Munasinghe, *Optimization and Optimal Control in a Nutshell*, Engineering Optimization: Methods and Applications,
https://doi.org/10.1007/978-981-97-8167-6

Appendix E
Matlab Code 6.1

```
% Free final time, Fixed final state OPtimal Control
clc; clear LL; close all;

% 1. Declare state, co-state and control variables
syms x1(t) x2(t) Lam1(t) Lam2(t) u(t)
eqns = [diff(Lam1,t) == -x1, ...
        diff(Lam2,t) == -3*x2-Lam1+2*Lam2, ...
        diff(x1,t) == x2,...
        diff(x2,t) == -2*x2-0.5*Lam2];
% 2. Solve State and co-state first order ODEs
S=dsolve(eqns); % derive symbolic expressions
S.u = -0.5*S.Lam2; % Define u

%3. Introduce boundary conditions
    syms t0 tf
    t0 = 0;
    eqns = [subs(S.x1,t0) == 0,...
            subs(S.x2,t0) == 1,...
            subs(S.x1,tf) == 3,...
            subs(S.x2,tf) == 5,...
            0.5*(subs(S.x1,tf)^2+3*subs(S.x2,tf)^2+2*subs(S.u,tf)^2)+...
            subs(S.Lam1,tf)*subs(S.x2,tf)+subs(S.Lam2,tf)*(-2*subs(S.x2,tf)...
            +subs(S.u,tf))==0];

% Determine ODE coefficients C1,C2,C3,C4
    SS = solve(eqns); % determine coefficients
    C1 = SS.C1;
    C2 = SS.C2;
    C3 = SS.C3;
    C4 = SS.C4;
    tf = SS.tf;

% Plot Results
points = 20; t=t0:tf/points:tf; % set 20 points
x1 = double(subs(S.x1));
x2 = double(subs(S.x2));
Lam1 = double(subs(S.Lam1));
Lam2 = double(subs(S.Lam2));
u = double(subs(S.u));

subplot(121);plot(t,x1,'k+-',t,x2,'ko-',t,u,'k*-'); legend('$x_1$',...
    '$x_2$','$u$','interpreter','latex'); grid on
xlabel('Time[s]');
subplot(122);plot(t,Lam1,'k+-',t,Lam2,'ko-'); legend('$\lambda1$',...
    '$\lambda2$','interpreter','latex'); grid on
xlabel('Time[s]');
```


S. R. Munasinghe, *Optimization and Optimal Control in a Nutshell*, Engineering Optimization: Methods and Applications,
https://doi.org/10.1007/978-981-97-8167-6

Appendix F
Matlab Code 6.2.1

```
% Fixed final time, and free final state Optimal Control
clc; clear all; close all;

% 1. Declare state, co-state and control variables
syms x1(t) x2(t) Lam1(t) Lam2(t) u(t)
eqns = [diff(Lam1,t) == -x1, ...
        diff(Lam2,t) == -3*x2-Lam1+2*Lam2, ...
        diff(x1,t) == x2,...
        diff(x2,t) == -2*x2-0.5*Lam2];
% 2. Solve State and co-state first order ODEs
S=dsolve(eqns); % derive symbolic expressions
S.u = -0.5*S.Lam2; % Define u

%3. Introduce boundary conditions
    syms t0 tf
    t0 = 0; tf=2;
    eqns = [subs(S.x1,t0) == 0,...
            subs(S.x2,t0) == 1,...
            subs(S.x1,tf) == 5,...
            (subs(S.x2,tf)-5)-subs(S.Lam2,tf) ==0];

% Determine ODE coefficients C1,C2,C3,C4
    SS = solve(eqns); % determine coefficients
    C1 = SS.C1;
    C2 = SS.C2;
    C3 = SS.C3;
    C4 = SS.C4;

% Plot Results
points = 20; t=t0:tf/points:tf; % set 20 points
x1 = double(subs(S.x1));
x2 = double(subs(S.x2));
Lam1 = double(subs(S.Lam1));
Lam2 = double(subs(S.Lam2));
u = double(subs(S.u));

subplot(121);plot(t,x1,'k+-',t,x2,'ko-',t,u,'k*-'); legend('$x_1$',...
    '$x_2$','$u$','interpreter','latex'); grid on
xlabel('Time[s]');
subplot(122);plot(t,Lam1,'k+-',t,Lam2,'ko-'); legend('$\lambda1$',...
    '$\lambda2$','interpreter','latex'); grid on
xlabel('Time[s]');
```


S. R. Munasinghe, *Optimization and Optimal Control in a Nutshell*, Engineering Optimization: Methods and Applications,
https://doi.org/10.1007/978-981-97-8167-6

Appendix G
Matlab Code 6.2.1

```
% Fixed final time, and Fixed final state Optimal Control
clc; clear all; close all;

% 1. Declare state, co-state and control variables
syms x1(t) x2(t) Lam1(t) Lam2(t) u(t)
eqns = [diff(Lam1,t) == -x1, ...
        diff(Lam2,t) == -3*x2-Lam1+2*Lam2, ...
        diff(x1,t) == x2,...
        diff(x2,t) == -2*x2-0.5*Lam2];
% 2. Solve State and co-state first order ODEs
S=dsolve(eqns); % derive symbolic expressions
S.u = -0.5*S.Lam2; % Define u

%3. Introduce boundary conditions
    syms t0 tf
    t0 = 0; tf=3;
    eqns = [subs(S.x1,t0) == 1,...
            subs(S.x2,t0) == 1,...
            subs(S.x1,tf) == 6,...
            subs(S.x2,tf) == 3];

% Determine ODE coefficients C1,C2,C3,C4
    SS = solve(eqns); % determine coefficients
    C1 = SS.C1;
    C2 = SS.C2;
    C3 = SS.C3;
    C4 = SS.C4;

% Plot Results
points = 20; t=t0:tf/points:tf; % set 20 points
x1 = double(subs(S.x1));
x2 = double(subs(S.x2));
Lam1 = double(subs(S.Lam1));
Lam2 = double(subs(S.Lam2));
u = double(subs(S.u));

subplot(121);plot(t,x1,'k+-',t,x2,'ko-',t,u,'k*-'); legend('$x_1$',...
    '$x_2$','$u$','interpreter','latex'); grid on
xlabel('Time[s]');
subplot(122);plot(t,Lam1,'k+-',t,Lam2,'ko-'); legend('$\lambda1$',...
    '$\lambda2$','interpreter','latex'); grid on
xlabel('Time[s]');
```


S. R. Munasinghe, *Optimization and Optimal Control in a Nutshell*, Engineering Optimization: Methods and Applications,
https://doi.org/10.1007/978-981-97-8167-6

Appendix H
Matlab Code 7.1.1

```
% LQR with fixed tf and fixed xf
clc; clear all; close all;

A = [0 1;-1 2];
B = [0;1];
Q = [1 2;2 3];
R = 0.5;
Ptf = [1 1;1 3];
x0 = [2;-1];
t0 = 0; tf = 8;

% 1. Solve Matrix Riccati Equation from tf-to-t0
[t, P] = ode45(@(t,P)MDE(t, P, A, B, Q, R), [tf t0], Ptf);
% P is a row vector
t = t(end:-1:1); % create forward time line
PP = P(end:-1:1,:); % PP(t0) to PP(tf) as rows

% 2. Optimal state
xx = zeros(length(t),length(x0)); % create row-wise state column vector
xx(1,:) = x0;
for i=2:1:length(t)
    P = reshape(PP(i-1,:),size(A));
    [tt,x] = ode45(@(t,x)xdotfun(t,x,P,A,B,Q,R),[t(i-1) t(i)],xx(i-1,:)');
    xx(i,:) = x(end,:);
end

% 3. Optimal control
uu = zeros(length(t),size(B,2));
for i=1:1:length(t)
    P = reshape(PP(i,:),size(A));
    x = xx(i,:)';
    uu(i,:) = -(1/R)*B'*P*x;
end

% Plotting
u = uu; x1 = xx(:,1); x2 = xx(:,2);
P11 = PP(:,1); P21 = PP(:,2); P12 = PP(:,3); P22 = PP(:,4);

subplot(121);
plot(t,x1,'k+-',t,x2,'ko-',t,u,'k*-'); legend('$x_1$',...
    '$x_2$','$u$','interpreter','latex'); grid on
xlabel('Time[s]');

subplot(122);
plot(t,P11,'k+-',t,P12,'ko-',t,P21,'k*-',t,P22,'k-');
```


S. R. Munasinghe, *Optimization and Optimal Control in a Nutshell*, Engineering Optimization: Methods and Applications,
https://doi.org/10.1007/978-981-97-8167-6

```
legend('$P_{11}$','$P_{12}$','$P_{21}$','$P_{22}$','interpreter',...
    'latex'); grid on
xlabel('Time[s]');

function Pdot = MDE(t, P, A, B, Q, R)
    P = reshape(P,size(A)); % convert raw P to matrix P
    Pdot = -P*A-A'*P-Q+P*B*(1/R)*B'*P; % rate of chnage of P
    Pdot = Pdot(:); % Append second columns to first column
end

function xdot = xdotfun(t, x, P, A, B, Q, R)
    xdot = (A-B*(1/R)*B'*P)*x;
end
```

Appendix I
Matlab Code 7.2.1

```
% Infinite-time LQR and PID comparison

clc; clear; close all;

A = [0 1;-1 2];
B = [0;1];
x0 = [2;-1];
x0pid = x0(1);

% for infinite-time LQR
C = [1 0;0 1]; % read both states
D = [0;0];

% for PID COntrol
Cpid = [1 0]; % read only the first state
Dpid = 0;
Umax = 20;
Gs = ss(A,B,Cpid,Dpid);
Cs = pidtune(Gs,'pid')
% auto tuned PID gains are too high
% lower PID gains to LQR comparable response
Cs.Kp = 8
Cs.Ki = 2
Cs.Kd = 8

% Run both LQR and PID control
sim('it_LQR')
t = ans.ScopeData.time;

% LQR Result
x1 = ans.ScopeData.signals(1).values(:,1);
x2 = ans.ScopeData.signals(1).values(:,2);
u = ans.ScopeData.signals(2).values(:,1);

% PID Result
x1pid = ans.ScopeData1.signals(1).values(:,1);
upid = ans.ScopeData1.signals(2).values(:,1);

subplot(121);plot(t,x1,'k-',t,x1pid,'k--'); legend('$x_{1LQR}$',...
    '$x_{1PID}$','interpreter','latex'); grid on
xlabel('Time[s]'); yaxis([-8 8]);
subplot(122);plot(t,u,'k-',t,upid,'k--'); legend('$u_{LQR}$',...
    '$u_{PID}$','interpreter','latex'); grid on
xlabel('Time[s]'); yaxis([-8 8]); % yaxis([-Umax 10]);
```


S. R. Munasinghe, *Optimization and Optimal Control in a Nutshell*, Engineering Optimization: Methods and Applications,
https://doi.org/10.1007/978-981-97-8167-6

Appendix J
Matlab Code 9.2

```
% Discrete Linear Unconstrained LQR
% A 2nd order LTI State space model is discretized with 0.2s sampling
% The descrete SSM starts with x0 and regulated to (0,0)
% Optimal LQR used: u = -H\F*x0

clear; close all;
ts = 0.2; T=2; N = T/ts; % 10 steps - dont change as it affects Matrix sizes
t = [0:ts:T-ts];
n=2; m=1;
Ac = [0  1;
     -2 -1];
Bc = [0;
      1];
Cc = [1 0;
      0 1];
Dc = [0;
      0];
SSMc = ss(Ac,Bc,Cc,Dc);
SSMd = c2d(SSMc,ts);
A=SSMd.A; B=SSMd.B; C=SSMd.C; D=SSMd.D;

x0=[0.3;-0.3];
Q = [1 0;
     0 1];
r= 0.3;
z = zeros(2,1);
Qb = eye(2*N,2*N); Rb=r*eye(m*N,m*N);
Ab = [A;A^2;A^3;A^4;A^5;A^6;A^7;A^8;A^9;A^10];
Cb = [B     z     z     z     z     z     z     z     z     z;
     A*B   B     z     z     z     z     z     z     z     z;
     A^2*B A*B   B     z     z     z     z     z     z     z;
     A^3*B A^2*B A*B   B     z     z     z     z     z     z;
     A^4*B A^3*B A^2*B A*B   B     z     z     z     z     z;
     A^5*B A^4*B A^3*B A^2*B A*B   B     z     z     z     z;
     A^6*B A^5*B A^4*B A^3*B A^2*B A*B   B     z     z     z;
     A^7*B A^6*B A^5*B A^4*B A^3*B A^2*B A*B   B     z     z;
     A^8*B A^7*B A^6*B A^5*B A^4*B A^3*B A^2*B A*B   B     z;
     A^9*B A^8*B A^7*B A^6*B A^5*B A^4*B A^3*B A^2*B A*B   B];
F = transpose(Ab'*Qb*Cb);
H = Cb'*Qb*Cb + Rb;
u = -H\F*x0;

sim('DT_LQ.slx')
t = ans.ScopeData.time;
x1 = ans.ScopeData.signals(1).values(1,:)';
```


S. R. Munasinghe, *Optimization and Optimal Control in a Nutshell*, Engineering Optimization: Methods and Applications,
https://doi.org/10.1007/978-981-97-8167-6

```
x2 = ans.ScopeData.signals(1).values(2,:)';
u = ans.ScopeData.signals(2).values(1,:)';

plot(t,x1,'k-',t,x2,'k--',t,u,'ko'); grid on; hold on;
stairs(t,u); xaxis([0 2]); yaxis([-0.31 0.27]);
%stem(t,u)
legend('$x_1(k)$','$x_2(t)$','$u^*(k)$','interpreter','latex');
xlabel('k[0.2s steps]');

%lsim(SSMd,u',t,x0); grid on; xlabel('k'); ylabel('x_1,x_2, u');
```

Appendix K
Matlab Code 9.3

```
% Discrete Linear LQ optimal control (Regulation)
% Both unconstrained and constrained control sequences

% A 2nd order LTI State space model is discretized with 0.2s sampling
% interval. The descrete SSM starts with x0 and regulated to (0,0) using
% optimal control sequence u = -H\F*x0

clc; clear; close all;

ts = 0.2; T=2; N = T/ts; % 10 steps - dont change as it affects
t = [0:ts:T-ts];
n=2; m=1;
Ac = [0  1;
     -2 -1];
Bc = [0;
      1];
Cc = [1 0;
      0 1];
Dc = [0;0];
SSMc = ss(Ac,Bc,Cc,Dc);
SSMd = c2d(SSMc,ts);
A=SSMd.A; B=SSMd.B; C=SSMd.C; D=SSMd.D;

x0=[0.3;-0.3];
Q = [1 0;
     0 1];
r= 0.3;
z = zeros(2,1);
Qb= eye(2*N,2*N); Rb=r*eye(m*N,m*N);
Ab= [A;A^2;A^3;A^4;A^5;A^6;A^7;A^8;A^9;A^10];
Cb= [B      z      z      z      z      z      z      z      z    z;
     A*B    B      z      z      z      z      z      z      z    z;
     A^2*B  A*B    B      z      z      z      z      z      z    z;
     A^3*B  A^2*B  A*B    B      z      z      z      z      z    z;
     A^4*B  A^3*B  A^2*B  A*B    B      z      z      z      z    z;
     A^5*B  A^4*B  A^3*B  A^2*B  A*B    B      z      z      z    z;
     A^6*B  A^5*B  A^4*B  A^3*B  A^2*B  A*B    B      z      z    z;
     A^7*B  A^6*B  A^5*B  A^4*B  A^3*B  A^2*B  A*B    B      z    z;
     A^8*B  A^7*B  A^6*B  A^5*B  A^4*B  A^3*B  A^2*B  A*B    B    z;
     A^9*B  A^8*B  A^7*B  A^6*B  A^5*B  A^4*B  A^3*B  A^2*B  A*B  B];
F = transpose(Ab'*Qb*Cb);
H = Cb'*Qb*Cb + Rb;
u = -H\F*x0;

% Unconstrained Optimal control sequence
sim('DT_LQ.slx')
t = ans.ScopeData.time;
x1 = ans.ScopeData.signals(1).values(1,:)';
x2 = ans.ScopeData.signals(1).values(2,:)';
```


S. R. Munasinghe, *Optimization and Optimal Control in a Nutshell*, Engineering Optimization: Methods and Applications,
https://doi.org/10.1007/978-981-97-8167-6

```
u = ans.ScopeData.signals(2).values(1,:)';

subplot(121);plot(t,x1,'k-',t,x2,'k--',t,u,'ko'); grid on; hold on;
stairs(t,u); xaxis([0 5]); yaxis([-0.31 0.27]);
legend('$x_1$','$x_2$','$u$','interpreter','latex');
xlabel('k[0.2s steps]');

% Constrained Optimal control sequence
rb = F*x0
LB = [-0.2;-0.2;-0.2;-0.2;-0.2;-0.2;-0.2;-0.2;-0.2;-0.2];
UB = [0.15;0.15;0.15;0.15;0.15;0.15;0.15;0.15;0.15;0.15];
Aine = []; bine = []; Aeq = []; beq = [];
[u,J] = quadprog(H,rb,Aine,bine,Aeq,beq,LB,UB);

sim('DT_LQ.slx')
t = ans.ScopeData.time;
x1 = ans.ScopeData.signals(1).values(1,:)';
x2 = ans.ScopeData.signals(1).values(2,:)';
u = ans.ScopeData.signals(2).values(1,:)';

subplot(122);plot(t,x1,'k-',t,x2,'k--',t,u,'ko'); grid on; hold on;
stairs(t,u); xaxis([0 5]); yaxis([-0.31 0.27]);
legend('$x_1$','$x_2$','$u$','interpreter','latex');
xlabel('k[0.2s steps]');
```

Appendix L
Matlab Code 10.2.3

```
% Model predictive Discrete Linear Qudratic Optimal Regulator
% with unbound and bound control sequences
% A 2nd order LTI State space model is discretized with 0.2s sampling
% The descrete SSM starts with x0 and regulated to (0,0) using
% optimal control sequence u = -H\F*x0
clc; clear all; close all;

ts = 0.2; Nh = 30; Dur = ts*Nh; % Sampling interval and simulation duration
tu = [0:ts:Dur-ts]; tx = [0:ts:Dur]; % time axes for u and x
N=10; % prediction Horizon
n=2; m=1; % Two state variables, and one conrol input
Ac = [0  1;
     -2 -1];
Bc = [0;
      1];
Cc = [1 0;
      0 1];
Dc = [0;0];
SSMc = ss(Ac,Bc,Cc,Dc);      % CT state space model
SSMd = c2d(SSMc,ts);         % DT state space model
A=SSMd.A; B=SSMd.B; C=SSMd.C; D=SSMd.D;

x0=[0.3;-0.3];
Q = [1 0;
     0 1];
r= 0.3; % control signal is less costly
z = zeros(2,1);
Qb= eye(2*N,2*N); Rb=r*eye(m*N,m*N);
Ab= [A;A^2;A^3;A^4;A^5;A^6;A^7;A^8;A^9;A^10];
Cb= [B     z     z     z     z     z     z     z     z   z;
     A*B   B     z     z     z     z     z     z     z   z;
     A^2*B A*B   B     z     z     z     z     z     z   z;
     A^3*B A^2*B A*B   B     z     z     z     z     z   z;
     A^4*B A^3*B A^2*B A*B   B     z     z     z     z   z;
     A^5*B A^4*B A^3*B A^2*B A*B   B     z     z     z   z;
     A^6*B A^5*B A^4*B A^3*B A^2*B A*B   B     z     z   z;
     A^7*B A^6*B A^5*B A^4*B A^3*B A^2*B A*B   B     z   z;
     A^8*B A^7*B A^6*B A^5*B A^4*B A^3*B A^2*B A*B   B   z;
     A^9*B A^8*B A^7*B A^6*B A^5*B A^4*B A^3*B A^2*B A*B B];
F = transpose(Ab'*Qb*Cb);
H = Cb'*Qb*Cb + Rb;
C1 = [1 0 0 0 0 0 0 0 0 0];
x(:,1)=x0;
for k = 1:Nh
    u(1,k) = -C1*(H\F*x(:,k));
    x(:,k+1) = A*x(:,k)+B*u(1,k);
end

subplot(121);plot(tx,x(1,:),'k-',tx,x(2,:),'k--',tu,u,'ko'); grid on; hold on;
```


S. R. Munasinghe, *Optimization and Optimal Control in a Nutshell*, Engineering Optimization: Methods and Applications,
https://doi.org/10.1007/978-981-97-8167-6

```
stairs(tu,u); xaxis([0 5]); yaxis([-0.31 0.27]);
legend('$x_1$','$x_2$','$u$','interpreter','latex');
xlabel('k[0.2s steps]');

% Constrained Optimal control sequence
LB = [-0.2;-0.2;-0.2;-0.2;-0.2;-0.2;-0.2;-0.2;-0.2;-0.2];
UB = [0.2;0.2;0.2;0.2;0.2;0.2;0.2;0.2;0.2;0.2];
Aine = []; bine = []; Aeq = []; beq = [];
rb = F*x0
[U,J] = quadprog(H,rb,Aine,bine,Aeq,beq,LB,UB);

for k = 1:Nh
    [U,J] = quadprog(H,rb,Aine,bine,Aeq,beq,LB,UB);
    u(1,k) = C1*U;
    x(:,k+1) = A*x(:,k)+B*u(1,k);
    rb = F*x(:,k+1);
end

subplot(122);plot(tx,x(1,:),'k-',tx,x(2,:),'k--',tu,u,'ko'); grid on; hold on;
stairs(tu,u); xaxis([0 5]); yaxis([-0.31 0.27]);
legend('$x_1$','$x_2$','$u$','interpreter','latex');
xlabel('k[0.2s steps]');
```

Appendix M
Matlab Code 10.3.2

```
% Model predictive Discrete Linear Qudratic Optimal Tracking Control
% with unbound control signals
% A 2nd order LTI State space model is discretized with 0.2s sampling

clear all; close all;

ts = 0.2; Nh = 40; Dur = ts*Nh; % Sampling interval and simulation duration
tu = [0:ts:Dur-ts]; tx = [0:ts:Dur]; % 1-step offset time axes for u and x

% Reference Trajectory
Dis = 1.6; Acc = Dis/(Dur/2)^2;
r(1) = 0; v(1) =0;
for k=1:Nh/2
    r(k+1) = 0.5*Acc*(ts*k)^2;
    v(k+1) = Acc*(ts*k);
end
for k=1:Nh/2
    r(Nh/2+k+1) = r(Nh/2+1)+v(Nh/2+1)*(ts*k)-0.5*Acc*(ts*k)^2;
    v(Nh/2+k+1) = v(Nh/2+1)-Acc*(ts*k);
end

for k=1:10
    r(Nh+k+1) = r(Nh+1); % add 10 more reference values beyond Nh
end

n=2; m=1; % Two state variables, and one conrol input
Ac = [0 1;
     -2 -1];
Bc = [0;
      1];
Cc = [1 0];
Dc = 0;
SSMc = ss(Ac,Bc,Cc,Dc);       % CT state space model
SSMd = c2d(SSMc,ts);          % DT state space model
A=SSMd.A; B=SSMd.B; C=SSMd.C; D=SSMd.D;

x0=[0;0];
Q = 1; QN = 3;
R = 0.3; % control signal is less costly
z21 = zeros(2,1);
Ab = [A B;
      z21' 1];
Bb = [B;
      1];
Cb = [C 0];
```


S. R. Munasinghe, *Optimization and Optimal Control in a Nutshell*, Engineering Optimization: Methods and Applications,
https://doi.org/10.1007/978-981-97-8167-6

```
q = Cb'*Q*Cb; qN = Cb'*QN*Cb; z33=zeros(3,3)
Qbb = [ q z33 z33 z33 z33 z33 z33 z33 z33 z33;
        z33 q  z33 z33 z33 z33 z33 z33 z33 z33;
        z33 z33 q  z33 z33 z33 z33 z33 z33 z33;
        z33 z33  z33 q z33 z33 z33 z33 z33 z33;
        z33 z33  z33 z33 q z33 z33 z33 z33 z33;
        z33 z33  z33 z33 z33 q z33 z33 z33 z33;
        z33 z33  z33 z33 z33 z33 q z33 z33 z33;
        z33 z33  z33 z33 z33 z33 z33 q z33 z33;
        z33 z33  z33 z33 z33 z33 z33 z33 q z33;
        z33 z33  z33 z33 z33 z33 z33 z33 z33 qN;]
z13 = zeros(1,3); QCb=Q*Cb; QNCb=QN*Cb;
Tbb = [QCb z13 z13 z13 z13 z13 z13 z13 z13 z13;
        z13 QCb z13 z13 z13 z13 z13 z13 z13 z13;
        z13 z13 QCb z13 z13 z13 z13 z13 z13 z13;
        z13 z13 z13 QCb z13 z13 z13 z13 z13 z13;
        z13 z13 z13 z13 QCb z13 z13 z13 z13 z13;
        z13 z13 z13 z13 z13 QCb z13 z13 z13 z13;
        z13 z13 z13 z13 z13 z13 QCb z13 z13 z13;
        z13 z13 z13 z13 z13 z13 z13 QCb z13 z13;
        z13 z13 z13 z13 z13 z13 z13 z13 QCb z13;
        z13 z13 z13 z13 z13 z13 z13 z13 z13 QNCb;];

Rbb = [R 0 0 0 0 0 0 0 0 0;
        0 R 0 0 0 0 0 0 0 0;
        0 0 R 0 0 0 0 0 0 0;
        0 0 0 R 0 0 0 0 0 0;
        0 0 0 0 R 0 0 0 0 0;
        0 0 0 0 0 R 0 0 0 0;
        0 0 0 0 0 0 R 0 0 0;
        0 0 0 0 0 0 0 R 0 0;
        0 0 0 0 0 0 0 0 R 0;
        0 0 0 0 0 0 0 0 0 R;]

Abb = [Ab;Ab^2;Ab^3;Ab^4;Ab^5;Ab^6;Ab^7;Ab^8;Ab^9;Ab^10];
z31 = zeros(3,1);
Bbb=[...
Bb       z31      z31      z31      z31      z31      z31      z31      z31    z31;
Ab*Bb    Bb       z31      z31      z31      z31      z31      z31      z31    z31;
Ab^2*Bb  Ab*Bb    Bb       z31      z31      z31      z31      z31      z31    z31;
Ab^3*Bb  Ab^2*Bb  Ab*Bb    Bb       z31      z31      z31      z31      z31    z31;
Ab^4*Bb  Ab^3*Bb  Ab^2*Bb  Ab*Bb    Bb       z31      z31      z31      z31    z31;
Ab^5*Bb  Ab^4*Bb  Ab^3*Bb  Ab^2*Bb  Ab*Bb    Bb       z31      z31      z31    z31;
Ab^6*Bb  Ab^5*Bb  Ab^4*Bb  Ab^3*Bb  Ab^2*Bb  Ab*Bb    Bb       z31      z31    z31;
Ab^7*Bb  Ab^6*Bb  Ab^5*Bb  Ab^4*Bb  Ab^3*Bb  Ab^2*Bb  Ab*Bb    Bb       z31    z31;
Ab^8*Bb  Ab^7*Bb  Ab^6*Bb  Ab^5*Bb  Ab^4*Bb  Ab^3*Bb  Ab^2*Bb  Ab*Bb    Bb     z31;
Ab^9*Bb  Ab^8*Bb  Ab^7*Bb  Ab^6*Bb  Ab^5*Bb  Ab^4*Bb  Ab^3*Bb  Ab^2*Bb  Ab*Bb  Bb];

Hbb = Bbb'*Qbb*Bbb + Rbb;
FbbT =[Abb'*Qbb*Bbb;
       -Tbb*Bbb];
Fbb = FbbT';

C1 = [1 0 0 0 0 0 0 0 0 0];
x(:,1) = x0;
up = 0; % u(k-1)

for k = 1:Nh
    xb(:,k) = [x(:,k);up;r(k:9+k)'];
    Du(k) = -C1*(Hbb\Fbb*xb(:,k));
    u(k) = up + Du(k);
    x(:,k+1) = A*x(:,k)+B*u(k);
    up = u(k);
end

plot(tx,r(1:41),'k--',tx,x(1,:),'k-',tx,x(2,:),'k*',tu,u,'o'); grid on; hold on;
stairs(tu,u); %xaxis([0 5]); yaxis([-0.31 0.27]);
legend('$r$','$x_1$','$x_2$','$u$','interpreter','latex');
xlabel('k[0.2s steps]');
```

Appendix N
Matlab Code 10.3.2

```
% Model predictive Discrete Linear Qudratic Optimal Tracking Control
% with control limits
% A 2nd order LTI State space model is discretized with 0.2s sampling

clear all; close all;

ts = 0.2; Nh = 40; Dur = ts*Nh; % Sampling interval and simulation duration
tu = [0:ts:Dur-ts]; tx = [0:ts:Dur]; % 1-step offset time axes for u and x

% Reference Trajectory
Dis = 1.6; Acc = Dis/(Dur/2)^2;
r(1) = 0; v(1) =0;
for k=1:Nh/2
    r(k+1) = 0.5*Acc*(ts*k)^2;
    v(k+1) = Acc*(ts*k);
end
for k=1:Nh/2
    r(Nh/2+k+1) = r(Nh/2+1)+v(Nh/2+1)*(ts*k)-0.5*Acc*(ts*k)^2;
    v(Nh/2+k+1) = v(Nh/2+1)-Acc*(ts*k);
end

for k=1:10
    r(Nh+k+1) = r(Nh+1); % add 10 more reference values beyond Nh
end

n=2; m=1; % Two state variables, and one conrol input
Ac = [0 1;
      -2 -1];
Bc = [0;
       1];
Cc = [1 0];
Dc = 0;
SSMc = ss(Ac,Bc,Cc,Dc);       % CT state space model
SSMd = c2d(SSMc,ts);          % DT state space model
A=SSMd.A; B=SSMd.B; C=SSMd.C; D=SSMd.D;

x0=[0;0];
Q = 1; QN = 3;
R = 0.3; % control signal is less costly
z21 = zeros(2,1);
Ab = [A B;
       z21' 1];
Bb = [B;
       1];
Cb = [C 0];

q = Cb'*Q*Cb; qN = Cb'*QN*Cb; z33=zeros(3,3)
```


S. R. Munasinghe, *Optimization and Optimal Control in a Nutshell*, Engineering Optimization: Methods and Applications,
https://doi.org/10.1007/978-981-97-8167-6

```
Qbb = [ q z33 z33 z33 z33 z33 z33 z33 z33 z33;
      z33 q  z33 z33 z33 z33 z33 z33 z33 z33;
      z33 z33 q  z33 z33 z33 z33 z33 z33 z33;
      z33 z33  z33 q z33 z33 z33 z33 z33 z33;
      z33 z33  z33 z33 q z33 z33 z33 z33 z33;
      z33 z33  z33 z33 z33 q z33 z33 z33 z33;
      z33 z33  z33 z33 z33 z33 q z33 z33 z33;
      z33 z33  z33 z33 z33 z33 z33 q z33 z33;
      z33 z33  z33 z33 z33 z33 z33 z33 q z33;
      z33 z33  z33 z33 z33 z33 z33 z33 z33 qN;]
z13 = zeros(1,3); QCb=Q*Cb; QNCb=QN*Cb;
Tbb = [QCb z13 z13 z13 z13 z13 z13 z13 z13 z13;
      z13 QCb z13 z13 z13 z13 z13 z13 z13 z13;
      z13 z13 QCb z13 z13 z13 z13 z13 z13 z13;
      z13 z13 z13 QCb z13 z13 z13 z13 z13 z13;
      z13 z13 z13 z13 QCb z13 z13 z13 z13 z13;
      z13 z13 z13 z13 z13 QCb z13 z13 z13 z13;
      z13 z13 z13 z13 z13 z13 QCb z13 z13 z13;
      z13 z13 z13 z13 z13 z13 z13 QCb z13 z13;
      z13 z13 z13 z13 z13 z13 z13 z13 QCb z13;
      z13 z13 z13 z13 z13 z13 z13 z13 z13 QNCb;];

Rbb = [R 0 0 0 0 0 0 0 0 0;
      0 R 0 0 0 0 0 0 0 0;
      0 0 R 0 0 0 0 0 0 0;
      0 0 0 R 0 0 0 0 0 0;
      0 0 0 0 R 0 0 0 0 0;
      0 0 0 0 0 R 0 0 0 0;
      0 0 0 0 0 0 R 0 0 0;
      0 0 0 0 0 0 0 R 0 0;
      0 0 0 0 0 0 0 0 R 0;
      0 0 0 0 0 0 0 0 0 R;]

Abb = [Ab;Ab^2;Ab^3;Ab^4;Ab^5;Ab^6;Ab^7;Ab^8;Ab^9;Ab^10];
z31 = zeros(3,1);
Bbb= [Bb       z31      z31      z31      z31      z31      z31      z31     z31;
     Ab*Bb    Bb       z31      z31      z31      z31      z31      z31     z31;
     Ab^2*Bb  Ab*Bb    Bb       z31      z31      z31      z31      z31     z31;
     Ab^3*Bb  Ab^2*Bb  Ab*Bb    Bb       z31      z31      z31      z31     z31;
     Ab^4*Bb  Ab^3*Bb  Ab^2*Bb  Ab*Bb    Bb       z31      z31      z31     z31;
     Ab^5*Bb  Ab^4*Bb  Ab^3*Bb  Ab^2*Bb  Ab*Bb    Bb       z31      z31     z31;
     Ab^6*Bb  Ab^5*Bb  Ab^4*Bb  Ab^3*Bb  Ab^2*Bb  Ab*Bb    Bb       z31     z31;
     Ab^7*Bb  Ab^6*Bb  Ab^5*Bb  Ab^4*Bb  Ab^3*Bb  Ab^2*Bb  Ab*Bb    Bb      z31;
     Ab^8*Bb  Ab^7*Bb  Ab^6*Bb  Ab^5*Bb  Ab^4*Bb  Ab^3*Bb  Ab^2*Bb  Ab*Bb   Bb;
     Ab^9*Bb  Ab^8*Bb  Ab^7*Bb  Ab^6*Bb  Ab^5*Bb  Ab^4*Bb  Ab^3*Bb  Ab^2*Bb Ab*Bb Bb];

Hbb = Bbb'*Qbb*Bbb + Rbb;
FbbT =[Abb'*Qbb*Bbb;
      -Tbb*Bbb];
Fbb = FbbT';

C1 = [1 0 0 0 0 0 0 0 0 0];
x(:,1) = x0;
up = 0; % u(k-1)

for k = 1:Nh
    xb(:,k) = [x(:,k);up;r(k:9+k)'];
    Du(k) = -C1*(Hbb\Fbb*xb(:,k));
    u(k) = up + Du(k);
    x(:,k+1) = A*x(:,k)+B*u(k);
    up = u(k);
end

subplot(121);
plot(tx,r(1:41),'k--',tx,x(1,:),'k-',tx,x(2,:),'k*',tu,u,'o'); grid on; hold on;
stairs(tu,u); yaxis([-0.2 3.3]);
legend('$r$','$x_1$','$x_2$','$u$','interpreter','latex');
xlabel('k[0.2s steps]');

% Constrained optimal control sequence
Umax = 5; Umin = -5; Umax1 = 2.5; Umin1 = -2.5;
LB = [Umin;Umin;Umin;Umin;Umin;Umin;Umin;Umin;Umin;Umin];
UB = [Umax;Umax;Umax;Umax;Umax;Umax;Umax;Umax;Umax;Umax];
Aine = []; bine = []; Aeq = []; beq = [];
x(:,1) = x0;; up = 0;
```

```
for k = 1:Nh
    rbb = Fbb*[x(:,k);up;r(k:9+k)'];
    if up>0
        UB(1,1) = Umax1-up;
    end
    if up<0
        LB(1,1) = Umin1-up;
    end
    [U,J] = quadprog(Hbb,rbb,Aine,bine,Aeq,beq,LB,UB);
    Du(k) = C1*U;
    u(k) = up + Du(k);
    x(:,k+1) = A*x(:,k)+B*u(k);
    up = u(k);
end

subplot(122);
plot(tx,r(1:41),'k--',tx,x(1,:),'k-',tx,x(2,:),'k*',tu,u,'o'); grid on; hold on;
stairs(tu,u); yaxis([-0.2 3.3]);
legend('$r$','$x_1$','$x_2$','$u$','interpreter','latex');
xlabel('k[0.2s steps]');
```

SPRINGER NATURE

GPSR Compliance

The European Union's (EU) General Product Safety Regulation (GPSR) is a set of rules that requires consumer products to be safe and our obligations to ensure this.

If you have any concerns about our products, you can contact us on ProductSafety@springernature.com

In case Publisher is established outside the EU, the EU authorized representative is:

Springer Nature Customer Service Center GmbH
Europaplatz 3
69115 Heidelberg, Germany

The manufacturer's authorised representative in the EU is Springer Nature Customer Service Centre GmbH, Europaplatz 3, 69115 Heidelberg, Germany. If you have any concerns regarding our products, please contact ProductSafety@springernature.com

Printed and bound by CPI Group (UK) Ltd, Croydon, CR0 4YY
30/09/2026
02225308-0005